I0505731

U.S. Army Ordnance Guided Missile School Badge

Bill Hewlett & Dave Packard Circa 1960s on Page Mill Road HQ Palo Alto, CA

NASA Space Shuttle (STS) Launch Cape Kennedy, FL...All Systems GO!

US ARMY Nike Hercules Missile – 1960 - Full up display and ready to go!

Preface
My 50 Years in High Technology

*My stories begin with my upbringing, a foundation for my decision making that was later to help me survive and thrive in the ever-increasing difficult jobs as an electronics technician, engineer, manager and director for World Class leading industry producers. And later, being a hands-on, working Manager for leading government contractors to the then best agency of the USA Federal government... **The National Aeronautics & Space Administration (NASA)**...one team consisting of the planet's best engineers...that began and was molded by the WWII winners, our **Greatest Generation***. The results of establishing the new Agency, in 1958, was to pioneer and arrive at man's most difficult journey...to SPACE and beyond, doing feats unheard of since the beginning of mankind...like reaching the moon in a decade and seeing it on Earth's television from one's own living room, all before thought of, as impossible!*

Our USA core values were derived and driven by this Depression Era raised, Greatest Generation...that everything was possible! And American Industry, NASA and its contractors, again and again proved, that it was true. If you can think it, you can do it, was embedded into each high technology team member's mindset.

From my early years, my generation was driven by our American culture to always strive to be winners.

We trained and played for it in sports and in our chosen fields; we were focused on being the best, whatever our endeavor. The USA beat the evil, bad guys in WWII and our Father's came home heroes of that magnificent world conflict.

At both NASA and HP, more than elsewhere, this winning attitude was shown in all our plans, missions and endeavors. It was quite a fascinating thing to see, be part of and experience...a team effort that overcame all odds and achieved mission success, again and again.

However, success was no accident, people needed the skills, experience and dedication and leadership, to attain the goals of a project or company. This book takes you through the rigors of the industry culture and NASA Project Management missions. The book tells and reveals the inside stories of how success is achieved and provides you, the reader, with a first-hand look at how our high technology business runs. Read on to see how the best agency in our Federal Government was successfully managed. And how they both keep on ticking, just like the watch commercial, taking us to new, scientific breakthroughs and hard to get destinations with the apparent ease of a walk through the park on a Sunday morning!

Book title **"Greatest Generation" by Tom Brokaw, Random House 1998*

1. In the Beginning: My Coal Cracker Roots

For me, it all began in a small Northeastern Pennsylvania coal-mining town. I was an offspring to one of the town's first cross-cultural marriages between the second-generation Irish Americans and the new immigrant Italians. My Irish Great-Grandfather, escaping from the 1840s Irish potato famines, fought in the USA Civil War and, as I was told by my Dad, he served in both the US Army and US Navy! At the time, in the middle 1860s, a three-month enlistment paid $300 US dollars…in the Army. However, if one joined up with the Navy, a new enlistee would get $600…so he did both, in that order. His Son, My Irish Grandfather, Robert retired from the Pennsylvania Railroad (PRR) in the early 1950s. He started his career by completing Correspondence School training in steam engines and worked his way up to Supervisor of a large train depot. Later, he worked as a train conductor, when he got older. And Buffalo to Baltimore was one of his routes. Somehow, between train runs, he became the Father of 17 living offspring with two more lost during childbirth! His lovely wife Sarah, a Gold Star Mother, with roots from County Mayo, Ireland, lost her first Son, Robby, in WWI. She was quite an activist, now called suffragettes, championing the right to vote and prohibition well before the Federal Government passed laws. And what about her 17-living offspring, how did she ever cope with that size family. My three kids were enough for me to handle. It was an Irish tradition then to have large families. As my Dad told me, the older girls took care of the young ones and as soon as the boys were old enough, off to the Big Apple, NYC, they went. There one could always find work to do no matter how menial the task! And working on the railroad with their Father was available at the large train repair depot called Coxton Yards, where he was a Supervisor.

On my Italian-side or as I was later told, Sicilian-side, was Grandfather, Joseph, who became an entrepreneur after suffering a serious leg injury working in the mines from sunup to sundown! He came to this country at the beginning of the 20th Century, after many tragic events occurred in his beloved Sicily. Such catastrophes drove him to leave, first, a volcanic eruption that killed tens of thousands and, secondly, a Bank of Italy collapse where he lost most of his savings. In addition, later doing research, I found out that a thing called "the Plague" helped him also decide to leave his beloved homeland. In the USA, he started his own grocery store business after a coal mining accident almost took out his leg. And I'm told that he fed the local neighborhood well during the Depression in the 1930s. Most neighbors racked up very high bills since there were no credit cards in those days, only his handwritten books that kept score. I'm told that few ever repaid their debts to him, but he struggled to keep everyone fed…that was how it was done in those days; neighbors helped neighbors since all were leveled by the USA stock market crash of 1929. One of his neighbors heavily involved in the stock market and what they called "rich" at the time never recovered his sanity after losing all his investments in the big crash! I was told they had to lay off all their hired help as a result, but still had their mansion…very run down in the then later 1940s. I recall my Grand Mother, Paulina, still sending food to that family, even after all those years passed. My Mother said they were the lucky ones to have a roof over their heads and food on the table. Such were the periods in the USA prior to what was called the big one, WWII!

Photo 1.0: Coal Miners - coming up from deep down under the Earth after day's work! Their next stop, happy-hour and boilermakers to wash down the coal dust! After getting courtesy hot showers on the mine's property.

My Sicilian Grandmother, whom everyone called Paulina, I considered a living saint. Very religious, she walked every Sunday to Mount Carmel's Church, miles away, up a high hill and at least 45 minutes to an hour from her home near the Susquehanna River. Rain, sleet or snow, nothing stopped her from attending mass at her church. My Grandfather, Joe, needed to go back to Sicily by boat from NYC three times to convince her Father to let him marry her and take her back to his adopted Pennsylvania. According to my Mother, her Mother was a key resource at her Grand Father's hotel in Sicily, in a place called Serra De Falco Province. She was the secret to his success as a culinary expert whose dishes brought people from miles away just to taste her famous homemade Italian specialties. When I was growing up nearby, most of my time was spent at her house where she treated me to dishes that I thought all Grandmothers cooked. You can imagine my first years of marriage, to someone who was a novice in the kitchen. For, I guess at the time, at 23, kitchen credentials where not what I had as a priority on my wish list! However today, my new wife, Julie, cooks in several languages and I tell my friends it took me 30 years to find her! My Sister, Peg, over the years I was single, told

3

me that I would never find a girl to meet my requirements. Peg was right, and I told her… not in the USA! I needed to go around the planet to China, to find her, a multi-talented dream girl, I was blessed to marry and never let her go away! All my friends say I was lucky to find her and who am I to disagree with that statement!

My Mom, Jerry, as all her friends called her, was very special to! She even worked at home later in life until she was 75 years old as a Notary Public. And she was also my Dad's Secretary when he was Justice of the Peace, later in his life. Mom was a very modern girl, seeing her 20s photos showed me why my Dad was smitten by her beauty. Her generation in the 1940s set the fashion bar high. All her friends were Hollywood beauties also. My Dad was one lucky Irishman to have her as his cheerleader all his life! Especially according to him, after their first weeks of marriage, he sent her home to learn to cook, after almost putting him in the hospital after one evening's dinner. She became a chip off her mother's block and always put on a great spread during family holidays! As I often said, she was never one to just sit around the house being bored. She always active socially, whether it was the local Woman's Club that she helped start and was a charter member or working to pay off their new home mortgage in 15 years, and that they did. In the 1980s, after my divorce, she organized all my bills and helped me stay out of bankruptcy. After sending her as much money as I could after two years, she said all my bills were paid off. What a relief that was, thank you Mom is all I can say. One of my most difficult times of my life was watching her slowly passing away, two weeks from 90, suffering from dementia and Alzheimer's. Up to her last day, she kept calling for her husband Charles, who she would never forget after almost 69 years of marriage. No doubt in my mind that they were truly the Greatest Generation ever! After their passing, Dad in 2005 and Mom in 2008, I miss their presence every day of my life that they helped make unforgettable!

Now let's talk about World War II, the 1940s and how those war years influenced me. As I think being alive in 1945, even at five years old, the big War did have deep impact on me and my mindset and my value system. During the War years since 1940, the year I was born, I evolved and adapted to a world at war. I remember collecting metal objects to take to the town junkyard with all my other childhood friends for supporting the war effort. At 5 and 6 years old, I recall us riding high in a large open back dump truck to the junkyard to collect our share of the proceeds. Perhaps 50 or 60 cents was my reward for doing such good deeds. The best games we played then, all involved war. Killing the Japs and the German Krauts was our main objective. I thought that war and being a soldier was as normal as the sun raising and setting and looked forward to the day, I too could be part of it. All my uncles and even some aunts were actively serving. My treasures were war souvenirs sent to me by relatives who were in the big battles. For at that time, war seemed to me to be normal and endless. All aspects of the World were in turmoil and had massive damage. Even the USA would suffer PTSD and psychological scars that are still healing today, such was the magnitude of emotional upheaval that WWII caused.

European families in the UK were also in a panic mode when both the German V1 and V2 rockets from Peenemunde launch sites were dropped randomly causing more terror than just damage and death. The problem was how to deal with this new technology that seemed to be unstoppable! No Patriot missile defenses were available then. Even in the USA, night air raid warnings were practiced. We turned all the lights out, even streetlights were dark too and all huddled around the radio to listen to the all-clear signal. The local Air Raid Warden usually knocked at our door to tell us it was over. I can only imagine the terror, especially in places like London, where the real war air raids were happening at nighttime daily.

Just walking around my neighborhood in those days resulted in unforgettable images. Each day, I would walk up the street to a local grocery store called Jimmy Lenza's. The streets were usually empty as most people were either in the service or working in defense plants or the mines. I still recall seeing many times one old World War I Vet, with his American Legion hat on singing each time he was in store "God Bless America!" Over and over, he kept singing! I even learned the words from him! And with a few pennies, I would make my sweet purchase. On my big days, I could afford a fresh ice cream cone, hand dipped by Jimmy, for five cents! This purchase was often a special treat that I experienced on the hot, summer days. Still not being able to read or write English, at five years old, I would look at the stands of daily newspaper headlines and see many large photos. One day, I vividly recall all papers had big headlines that said, "V E DAY" weeks later another headline "V J DAY" with photos of people celebrating on the foreign city streets and places I never even knew about. Later, one on April 12, 1945, had **"FDR DEAD!"** and a man named "HARRY S. TRUMAN" and his photo appeared. I knew that history was being made, as these newspaper Page 1 words were the headlines of the big World events for weeks to come.

I vividly recall, like it was yesterday, an Easter-time holiday in 1945, walking with my Uncle Pat, who was home on leave prior to going overseas to fight the evil Nazis. The ground at my grandmother's back yard was covered with daffodils in full bloom and I recall seeing myself in his spit-shined paratrooper boots holding dearly on his extended hand with a safe, secured feeling. Little did he know that in a few weeks he would wake up in a Paris hospital having been shot in both knees as he parachuted in combat jump number 10, over enemy lines. Pat is still alive in his nineties having lived 100% disabled all his life in the Southland most of his life and remarried at the age of 82! A 101st Airborne "Scream'in Eagle" WWII survivor hero in my eyes always!

For those lucky or unlucky for being alive during these war years, this was a historical time that would never happen again on the magnitude of this second world conflict. In 10 years starting in 1939, we had more technological breakthroughs than in the past thousand years and the following 10 years into 1959, even greater scientific breakthroughs than the last 10. Can you imagine experts today say, that in the next ten years, we will experience more technological advances than the last 100! We all have some good ones to look forward to…miracles are happening every single day!

My Dad, the second youngest out of several brothers, was a WWII Veteran, hero to me, he even served overseas in the Philippines…Manila and on the Island of Palawan nearby. My Dad did not need to serve. He had a family with offspring and could have been deferred; yet at he volunteered and became a Radio Repairman, who was saved by the two USA atomic bomb attacks in 1945. He instilled one driving force in me to "<u>be the best</u>" that is with me to this day! He, always my motivator (both positive and negative), later in his life, he asked me why I wasn't the head of NASA! I told him it was a very political job and he understood immediately what I meant. He loved politics and was heavily involved in local politics and in my estimation, would have made a great US Congressman or Senator had he taken the GI Bill college training after WWII. This was a difficult task for a family man with a wife and two kids to support, unlike me who also had a high paying job that helped us survive. At one time, I was taking 6 courses at three campuses and working a 56-hour workweek in the defense business, helping keep US Navy Top guns alive during Viet Nam!

In 1945, my Dad received Emergency Leave from his Fort Knox, KY Training Base. Back in PA, at a place called Number 9 Coal Mine, several blocks from his home, Dad's older brother Joe was suddenly killed in a coal mining accident. Joe had a large family with his two boys already in the military. His Son Billy was wounded in the WWII Battle of the Bulge and was awarded a Purple Heart from wounds suffered in that famous battle. I recall seeing this medal when he sent it home for safe keeping. It was unforgettable sight to see it; when it was taken out of its storage case, as we all were awed by such an honorable award. Joe's younger Son, Bobby, as we all called him, was in the US Navy. Uncle Joe as the story goes, was killed when trapped between two coal cars as they broke away from their highly-sloped railway.

Joe's wife Irene was one of my favorites as were all her offspring. She worked long hours to raise her large family of five girls and two boys. I recall both cousins, Billy's and Bobby's marriage ceremonies, one a Polish wedding reception to a beautiful girl named Dee, who my Dad introduced to Bobby. Dee later became a VP of Playtex and she designed several of their products during her successful career. His brother, Billy's wedding reception to another beauty named Naomi was especially memorable. At the close, all guests ended up singing "Good Night Irene" over and over almost like a Hollywood movie closing! "Irene Goodnight" was echoing even in the parking lot after the reception ended! Yes, that then famous hit, "Good Night Irene" I can still hear them singing that tune today! Both marriages yielded wonderful families who would make Uncle Joe and Aunt Irene very proud parents!

Back from the War II, I recall my Dad's welcome home party in nearby Hillside, New Jersey, where all his brothers and sisters settled to escape the mines and high unemployment of NE PA. From day one, I never liked NJ; perhaps it was because we always escaped back to PA every chance we would get. This also carried over to me when I too worked there. Jersey was a place to work, PA was a place to relax and enjoy our families and of course play too!

I was only 6 and remember a several days party that seemed to last forever, in his older brother's John's apartment in Hillside, NJ. The first night I still recall, with no place to sleep or bedclothes, I asked my Dad what to do? He replied, sleep on the floor like everyone else, which I tried to do. But I have no pillow Dad and he threw a couch pillow and said now get some sleep which somehow, I did! I recall seeing his brand-new Eisenhower Army jacket hanging over the kitchen chair with many decorations and one I never saw, he told me it was a ruptured duck for getting discharged from active duty! My Dad was my best friend in my life; we were like brothers when it came to work. I helped him complete electrical jobs he did all his life. I always raised hell, however, when he never charged folks for work, he did. Sometimes he collected just fees to cover his material that he got for his trade discount. Before I was 12, I learned to wire an entire house or put in a 220-volt electrical service. And, Dad was still doing jobs into his 80s, and if I were home, we did them together. I was always amazed of his strength; one tough dude he was

After the War, Dad opened-up a radio repair shop and it was filled with to be fixed radios that had a high failure rate with the new vacuum tube technology. And, for a few weeks I recall seeing a famous Wurlitzer floor model for repair, the one with a rainbow of colors glowing on each side, one that I still admire today. My first experience in repair was watching my Dad fix them. Soon, I too, followed his direction when he said, "replace that large black metal tube, the OZ4 and don't turn it on and get electrocuted!" I was delighted, at 6 years old, when he then would plug it in and it played like new, wow, I was working now just like my Dad I thought!

The store idea soon failed because radios were so cheap one could buy a new one for the price of a repair. Dad then went to his Plan B, a journey-man electrician, like his several brothers did their entire lives.

Dad was a great storyteller, as are most Irishmen! Two, I will share with you and they were his favorites for he loved to tell them repeatedly and we never tired of listening to them, it was great entertainment. The first was about his first week at Fort Knox, KY during his US ARMY Basic Training. He recalled being sent to the nearby railroad station to load heavily armored US Army tanks on railroad cars for shipment to Germany, as the Battle of the Bulge was underway, and we needed all the weapons we could gather for reversing the Nazi Ardennes Forest Offensive. My Dad, in the Army only a few weeks, said he thought it strange that he was selected for this task since he had no training on

9

driving tanks at all. So, he and other GIs somehow figured out what to do and loaded, all day long, all the tanks that was there. My Dad said he did about 15 to 20 himself. Well, the next morning, at roll call, his name was called out. Where were you yesterday? We have you as AWOL, said the Sargent in charge. My Dad replied that he was following orders and went to the railroad station and loaded tanks on RR cars all day long! The Sargent said WHAT! That was supposed to be Lt. Doyle, not my Dad, assigned to do that work, and everybody assembled had as good laugh! SNAFU, he added!

The next good one was when he first was married to my Mom, her Father told him that he had a new job for him and would show him what it was in the next morning. Before sunup, he and my Dad walked miles to the nearby mines where my Grandfather worked at the time. Miners in PA worked sun-up to sun-down! Only later, did a man named Mitchell (his statue is on the Courthouse Square downtown) fought for and won, in Scranton, after a Boston-like massacre riot occurred, did a 40-hour workweek started that, to this day, is the standard all over the USA. You can thank the coal miners in PA for that benefit. Back to my Dad's story. They both went down to the lower levels of the coalmine to the workplace. My Dad was told by my Grandfather, in his broken English, "Charlie we hava easy job today (the good news), only need to fill-up these 16 coal cars anda we go home!" My Dad said to himself not too bad a workday, however, then was told the bad news; the new coal is right around the corner! It took three loads to move the coal to the coal cars, quite a grueling job according to my Dad. During the process, my Dad said he heard a pinging noise and asked what it was? "Oh, youa new guys worry about every ting" my Grandfather replied. "Not a worry justa some coal cracken makin the noise from the pressure of the mine walls." My Dad did not buy that line either! After they were finished Dad said, "Thanks but no thanks, I'll stick to my electrical work", and he never went underground again…until he passed away. Two weeks later, what do you think? Yes, the mine ceilings where they were working just

10

collapsed. My Grandfather was lucky; the miners brought him home with just a broken leg (compound fracture!) and dropped him on his doorsteps of his home. If it was not for Dr. Dixon, who lived nearby, I guess he was just left to die at home as so many of them were back in those days. My Grandfather too, never went back underground either after that serous accident. He started a grocery business that he worked at all his life until killed in a car accident going to his brother's farm to pick fresh vegetables to sell at his store. It seemed a bee flew into the car and his brother panicked and hit a telephone pole and my Grandfather hit his head on the bullet-like dashboard compartment lock of a 1952 Ford that was demolished. His brother survived without a scratch! I recall visiting him daily in the local hospital for he never recovered consciousness from that serious head injury. It was something I never forgot, it took him 14 days, left almost untreated, to pass away. He was a hardworking, good man who everyone dearly loved and sorely missed.

I had many close friends growing up in my teen years. Many are still with us today as we now communicate on the world-wide web as they call it! All my friends were of European and Christian origin. Their parents or grandparents came from Italy, Ireland, Germany, England, Scotland, Poland and even Lithuania the regional makeup of the small community we all lived in. In my high school years, several fellow sports players even made it to the major leagues and like myself left our hometown to seek our own personal challenges. One I recall motivating me to continue to play football made it the second in command at IBM, he was a standout even then at an early age. Another fellow player became one of the largest contractors in the estate of Georgia and played with many NFL football stars back in the day. Local friends also were successful starting their own businesses. And all who could pass the draft physicals did their part serving in the US armed forces.

The city was filled with houses of worship; Catholic, Protestant, Orthodox all were part of every town's section from the so-called junction rail center to Browntown, Port

Griffith, Butler Hill, the Townships and Oregon on one side of the Susquehanna River to the West Side, Harding and the Falls, Wyoming, Exeter, Forty Fort on the other. All communities that made up of what they still call the "Valley with a Heart!"

I was approaching my 12th birthday and like all kids that age then, each night before I went to sleep prayed for a big gift for the future Christmas holiday…for it was six months away and surely plenty of time to start praying, hoping and wishing …for a new 26" bike, made in the USA! Well, months passed, and I recall my Dad had become a policeman filling in for another policeman who was drafted into the service to fight the war in Korea. Perhaps Dad, at that time, earned $50 dollars a week back then. And he was still paying on that new 1950 Chevy Deluxe Convertible, he bought after coming home from the war. Christmas time arrived and he said let's go…. you are going to get your Christmas present! He took me to Allen Smiles big bike store on Main Street were many new models were displayed. Perhaps there were 50 of them on display, all shinning and new! Pick one out, he said! I looked and looked and found mine; a new Columbia brand beauty, made in Massachusetts, USA and replied, Dad, here it is! Guess what, as he looked slightly startled at the price tag. He realized that I picked out the most expensive one in the store. It had a battery powered front light, brake light too and electric horn below the handlebars. And I loved the new color, sort of a spring shade, a brand new, tree leaf color of two-tone green. The only one in the store like it. And all my friends liked it too! Dad, I think ended up paying for it weekly as there were no credit cards around in those days. And Allen Smiles was just all smiles, as we rolled it out of the store. Don't forget it's guaranteed for 90 days and waved good luck with it! I loved that bike, but soon outgrew it, as my eyes were now on my Dad's Chevy convertible and getting my PA Driver's license. What happened to that bike you might be thinking? Don't remind me, it was like new until I left for the Army years later. Inquiring after returning home after Basic Training, I asked my Mother and she said, "Oh, we gave it to your cousin Vince with all your other paraphernalia and junk!" Including my baseball collection, I cried. "Yes, it is all gone, we thought you were never coming back!" My cards collection today would be worth a fortune not counting my bike that I saw on E-bay for over $7,000!

As for my childhood heroes, I guess they would start at my Dad. A football player in high school himself graduated in 1936 same year as my Mom. He dropped out two years to work in NYC like all his brothers in entry level jobs that paid little. Learning his lesson, he returned to PA and met my Mother, who was quite a catch back then. Her Father Joe, when asked for her hand in marriage, had only one qualifier, "Do you have a job, Charlie?" Yes! My Dad replied, "I'm an electrician working in New Jersey!" And prove it he did, for on his wedding day, he and Mom boarded an afternoon bus to Jersey, so my Dad could go to work that night! Mom loved to end that story with the following! Arriving in Jersey, Mom went to unpack his suitcase and found only a toothbrush! Where are all your clothes, she asked and he replied, "oh, they were all my brothers that I borrowed!" And they both had a good laugh! One day, I recall seeing my Dad in a new suit, wow, I thought, you sure have a handsome father! Dad had acquired quite a wardrobe once I was born. To his last days, he always dressed up like a Fifth Avenue, New Yorker, where he worked and took it, the NYC lifestyle, in as a teenager!

Other sports heroes name I recall today were held in high esteem! Names like Lombardi, Paterno, Joe Lewis not to forget those New York Champions like DiMaggio, Berra, Mantle, Maris, Campanella, Ford and Boston's Ted Williams! Local names like NFL star Charlie Trippi, who my Uncle Billy played with in high school were always talked about in our little community.

For some reason, President Truman was not held in high esteem like we did with Eisenhower, however, to me he was a great decision maker! Perhaps it was the TV, since I recall seeing IKE nominated by Gerald Ford in 1952, the first televised Republican Convention. Locally, we held our

13

Mayor, Superintendent of Schools and a few school Directors and Coaches, in high regard. Teachers in the late 1950s, had difficult jobs with our, rebels without a cause, generation. As is seen in many movies from that era, we wanted to be a different generation from our depression raised parents. I think we were more awakened to the world at the time. More informed than our parents by TV and radio, we saw the world through a different paradigm. Always suspicious and ever searching for life's answers, we were the first Americans to question society and our values, in general. What we first saw were the many injustices and inequalities of the world all around us. And we did not like what we saw. The few very rich verses the many poor, the suffering in our cities, the political corruption on a local, state and national level. The interrelationships between the so-called good guys and bad guys all around us. Yes, we were the rebels in our hearts for what we saw as a stacked deck world, we were soon to be part of and did not like what we witnessed one bit!

However, our Christian values helped to keep us grounded and harnessed the anger we held inside. Perhaps we could change the system by our newly granted voting powers? But then came the big changes over the next quarter a century...Viet Nam (1959-1975) and the Civil Rights Wars (1960-today). These were out of our control but affected us all individually even up to today. Viet Nam was to us, a completely unjustified war and all our major cities suffered protesting for justice with profound cries such as, "Hell no! We won't go!" and "Burn Baby Burn"!!

Once faced with these earth-shaking crises and coupled with our new adult responsibilities of marriage and child raising and the draft, our rebellious drives were soon subdued by the new pressures of earning a living and for tens of thousands of my generation serving in another war like our fathers "killing the yellow man" as Bruce Springsteen so correctly sang. We certainly grew up quickly during the 1960s and after JFK's Camelot World were faced with a just survive to stay alive defensive position...I think still haunting us to this day in this 21st Century. The 1960s flew by and soon the new political crises, one called Watergate consumed the land of the free and home of the brave!

One time much later, 1989 to be exact, as my Dad waited in his hospital room for a heart bypass operation in his 70s, his roommate was his high school friend supposedly connected to so-called high-level PA politicians. I could tell that upon nearing their room entrance, for the semi-private room looked like a NYC Deli filled lots of food and deserts, Italian delicacies, some even hanging from the ceiling! Dad introduced me and he motioned me to come closer. My Father's childhood friend held my hand tightly and asked what I did at NASA? And I told him, aerospace engineering. Are you an Astronaut, he replied? No, I replied, "I'm too old!" He looked at my face with a sincere grin like a scene out of the Godfather movie and said, "Do you want to be an Astronaut?" Supposedly, an offer I couldn't refuse. I'm probably too old, I told him. He grabbed my other hand and held my wrist and repeated "do you want to become an Astronaut?" Let me think about it, I replied! I looked at my Dad's face and he was laughing like hell! Later, he told me that if I wanted to, his friend could surely make it happen! The next day, we had my Dad transferred to a private room since he was not at a Club Med resort! He survived for 15 more years after a four-way bypass. His roommate survived his bypass, however, succumbed later in a Philadelphia hospital from secondary lung condition. The WWII folks were approaching old age and I did not like it!

My Dad's oldest brother, named Robby, a WWI US Army Veteran, never came home from overseas and succumbed at sea. His Mother, Sarah became a Goldstar member. This was an honor for all the USA WWI Mothers who lost their sons in the so-called War to end all Wars! She later toured Europe with other Gold Star Mothers who lost their sons courtesy of Uncle Sam. I always viewed photos, with my Dad, of her long trip by boat in the family photo album.

Only later, did I investigate about the places my Dad served as he seldom revealed the intimate details of the places and their history in the War. He did tell me that his buddies fought over the rope used to hang the Japanese General responsible for the Bataan Death March. Within 24 hours of his trial, he was quickly hung for his deeds. Dad was one of the lucky ones; he lived until almost 90 and had a prosperous life thanks to my devoted Mother and his workaholic disposition. He was saved by the big ones as he called them…Hiroshima and Nagasaki! I, only in recent years, discovered that what he termed a tropical paradise, Palawan Island! Dad's last post was nothing more at the time, than a former Japanese prison camp, where at the end of the War, a tragic atrocity occurred where hundreds of prisoners were slaughtered (burned alive in their bomb shelters) on orders of the retreating General. Only a few escaped, the book *__Last Man Out__* by Glenn McDowell, USMC covers the ordeal in vivid detail.

Today, Palawan Island is a tropical retreat, a wanting destination for weary workers looking for nirvana…an exotic vacation resort far off in the Pacific rim paradise of islands. I showed Dad it on the Internet before he passed away and he could not believe his eyes. "Been there done that" he replied, when I asked him if he would like to return! I was fortunate to have him in my life for 65 of my years…I still think and miss him every day, especially in Florida where he was lucky to escape those difficult Northern winters after he retired. He had a great saying about the Sunshine State, when during his last years, he stopped the winter snowbird retreats. "I'd rather freeze in Pennsylvania than freeze in Florida, at least I have a fireplace here!" So much for that great Florida public relations pitch! After several winters living there, I knew exactly what he means!

2. My Early Years: K 1-12 Kindergarten to Happy Days High School Years

My formal educational training began in 1945, at 5 years old, when I recall jumping daily into the big back seat of a new NASH 4-door automobile, with many other children, to be driven to the south side of town to a place called, Miss Smiles Kindergarten on Nafus Street. I recall this because in a few years, we would move to this same neighborhood, where I would develop some street smarts from all the Italian and Irish playmates from the area. Many of my new classmates continued to be in my classes' right through elementary, middle and junior and senior high school. Even at my future 50th Class Re-union most were still alive, and a lot remained in the town all their lives. But like me, those who left PA were far more ahead because we went on to colleges and worked in higher paid job locations throughout the USA.

My home in those early years was located close to a primary grade school called Fort Pittston. And from my home window each morning, I longed to join other kids who went to the school, past my home every day. Finally, at 6 years old, Mom told me to go join them and I was enrolled in the first grade, later to be a regret because I was much younger than my fellow students. All my school years, I suffered, especially in sports where fellow players were a few years bigger and stronger. However, I never let this be an obstacle and I performed with the best of them. After the third grade, my family moved to the south end of town, a melting pot of European American traditions that formed my political, cultural, religious and social values, most of them still today, part of my very being.

In the fourth grade, my teacher asked all class members what they wanted to be when they grew up…. me…nothing to do with coal mining…without a hitch, I replied "Aeronautical Engineering" having read about it in my treasured comic books at the time! All the class laughed by my reply…but they all stopped when they knew that I was serious and

17

explained what it entailed! Little did even I realize that my prognostication was going to be right on target!

Our town's education system was centered on Grades 1 to 6, 6 to 9, 9 to 12 with close neighborhood schools and teachers who taught me, some of whom, taught my parents. And all of them knew my parents from the many social activities scheduled mostly around the holidays. It was a great place to be raised. We were the winning generation after WWII. Our parents were all part of unified nation that overcame our enemies. The only negative was Korea, still today almost 60 years later, on our list of "World Hot Spots"! Ike was elected and a cease-fire was negotiated in 1953, to end the fighting, but not the hostilities. This was my first exposure of a conflict that we did not win, and I became confused by the unresolved conflict...for were we not the ones, the all-powerful USA, who saved the World as I was taught? Later, I would learn that many loose ends remained after the big one that today is still causing us new fears. And there are many nations in the world who are looking at the USA to help fix just about everything!

I graduated High School in June 1957, still at 16, a great year, with World at peace, right after Korea. With President Ike Eisenhower in charge, it became the foundation year for many 20th Century events and history, movies and black and white TV shows. After Sid Caesar and his Show of Shows on TV and Dragnet, one of my favorites was Amos and Andy. One can still buy those early TV episodes in Canada, but not in the USA, as Bill Crosby bought them, and they were never to be seen again in the USA. So much, I thought, for the so-called free speech movement.

Looking back on my high school courses, I was enrolled in the Academic Program, focused on courses that would be preparing us for college. All were required to take exclusively Latin as a language. Spanish was offered only to Commercial Program candidates, who were also to take typing and home economics. No cross course taking was allowed! In college, as well as my career, I could have used that Commercial training, but as males we were prohibited.

I also needed to learn to cook once I was back to being single years later! And on my many trips to Spanish speaking countries I could have used that language instead of a dead one called Latin! So much for my PA 50's school curriculum frozen in the 30's when my parents also graduated!

1957 was a big year for Space and high technology. The Russians became the first country to launch a 12-inch sphere with a beeping sound that shocked the planet! News media frenzies occurred, as repeatedly, this was the story of the year. In the USA, residents now had a new fear after living in the so-called Atomic Age. As duck and cover were daily exercises for kids in school all over the USA. As if it would be a protection for surviving nuclear attack. But I guess the government needed to do something to ease the fears we all had for many of the Cold War years. President Eisenhower knew better that we also had the technology to beat the Russians. Our military could have done the same thing as Sputnik; however, the plan was to use a Geo-Physical Year Celebration to launch our first satellite. A Redstone rocket would soon be launched with a more robust, higher technology, that even later included cameras that could be used around the globe to monitor Russian military bases and launch sites. This would be used to replace the ill-fated Lockheed U2 airplane that was finally hit by a missile while conducting a mission over the USSR. The Cold War was heating up, more than ever, even after, in 1953, when Joseph Stalin, Russian Dictator died, and hard liners took control of power in Communist Russia. I recall for years in my church praying for conversion of Russia and an end to post WWII tensions caused by our new enemy. This Cold War lasted the 1980s until President Ronald Reagan leadership and Star Wars technology almost bankrupted the entire USSR.

On September 11, 1950, I recall, at 10, marching alongside with the famous Pennsylvania National Guard, 109th Infantry troops in my hometown to the Exeter train station! Those poor soldiers never got to leave the USA, since their train was derailed in Ohio the next day, killing many of them in what I think was an act of sabotage to this day. Like all my new friends, I started playing football in the ninth grade, 1953 to be exact, at 13 years old. I wanted to be quarterback and needed to learn the playbook inside out for all players, since if they forgot what to do, I would tell them. After my second year, I learned the T-Formation and remembered my plays especially the passes, since I was a lefty and benefited from that trait. I only made the second string, since the guys I played with were two years older than me. Later, I appreciated that situation, when I attended my 50th Class Reunion and met the first stringer who had suffered serious back problems all his life from game injuries. I made up for it however in the Army, when I played for two years as a first stringer. In the 1950s, it was the decade that created top entertainment hits such as Happy Days, and movies like Rebel Without a Cause and Blackboard Jungle. Even today's hit TV show Mad Men still fascinates us and a whole new generation of 1950s fans. All of them reflected the mores of the times. And we all loved those triple years Chevy's, 1955, 6, 7. My favorite was the fuel-injected, 1957 black Bel Air Chevy 2 door hardtop with a continental kit on the back. And we all loved my good friend Frank Feeley's 1955 Mercury Montclair blue convertible, with Hollywood mufflers. Frank bought the car in California after taking a year off to work there. He later returned to LA, a s I was told later, and made it big time in real estate, in on the ground floor of the boom years. And, of course, I cannot forget to mention my sports heroes like Jolting Joe DiMaggio and Ted Williams and my Boston Braves now called the Milwaukee Braves who, that year, with Warren Span, Eddy Mathews and Hank Aaron, won the Pennant and World Series.

Photo 2.0: Hometown, Pittston, Pennsylvania Town mural …My Coal Cracker Roots!

Yes, now World Champions, the Braves, who I followed for years starting in Boston, where I dreamed one day of visiting, at least once in my lifetime! Ah yes! Boston, it seemed a little higher class than the local New York Yankees, Brooklyn Dodgers and NY Giants teams. And Babe Ruth, native Baltimorean, whose life storybook I read, played there for years, so to me it was on my future places to see hit list! Boston was a place that just fascinated me and acted as a lure from my teenage years. Little did I know, that I would spend almost 20 years of my life there going to college, teaching college, working in my chosen Electronics High Tech career and raising my family made up of a Long Island JAP, who in those days were referred to as a Jewish American Princess and three amazing kids.

In the 1950s, I kept a WWI Military inspired photo frame with a picture of General Douglas MacArthur on my bedroom dresser, along with all my boyhood treasures collection of baseball cards Ruth, Mantle, Ford, Campanella, Snyder, Mays, Rizzuto, Berra, etc.…hundreds of them! And I loved my new Columbia (US Massachusetts made) bicycle with the lighted taillight and cool sounding horn down from the handlebars. I rode that bike everywhere and all my friends envied my possessing it!

After high school, I intended to enroll to Penn State University. It was my Dad's desire that I become an electrical engineer! Little did I know the rigorous course work and mental preparation that college required? After two difficult years, at 17 and 18 years old, I flunked out! I had little preparation for college as my high school day's challenges were the prettiest girls and my red Chevy convertible, oh, and football, baseball and Rock and Roll too, not necessarily in that order either. The rigors, discipline and pre-college prerequisites were just not part of my mindset, I guess I could be called a late bloomer for later, I excelled in all my studies…but then I was married, I guess that did help me overcome my male priorities. After attending night school makeups for my now focused courses (English and Mechanics) I recall seeing all my friends enlisting in the service to "beat the draft" for a two-year active duty obligation. A USA government requirement at the time, was to serve active duty for two years or be trapped, drafted and be forced into a unit that was not your choice. Some of my more adventurous friends even joined the US Marines in what was described at the time was as a grueling ordeal and adventure!

I picked "pushing my draft"! A two-year Army hitch that could be gotten over quickly and easily…at least that is what I thought at 18 years old! And, as we all know, one knows all at 18, just ask someone who is in that age, a ready to vote now adult! Also, at 18, you did not need parental consent; when I told my Dad, he said thanks for the new birthday present as my time to leave for Fort Dix was March 9, 1959 and his birthday was March 7!

A catastrophic event occurred two weeks before I was to enter the service. This was to put a seal on ever doing the mines like my Grandfather once did! At around noontime, I was downtown with my friends when news came that the Susquehanna River that flowed through my hometown had come into the mines at Port Griffith, a mineshaft located near the riverbed. Unless you saw it, it was hard to describe. Think of a bathtub draining with a whirlpool of water going quickly down. Only this was the great river acting like a raging waterfall flowing into the roof of the mines. We rushed to the scene. I saw the elevator shaft struggling to come up to the surface and survivors on board. It appeared like 10 to 15 miners including one of my good friends Tommy Burn's Father among them. I also knew that my next-door neighbor Willy Sinclair, whose shinny 1950 black Ford 2 door coupe that he polished daily and I admired, was not among the group. We then walked up a long hill adjacent to the river and saw the disaster. A hugh, spiraling hole near the river's edge was sucking in water like a siphon into the roof of the mine! Many workers were trapped below, and the only survivors were thought to be what I saw on that elevator. For, try as they might, they could not get it to go down again. Panic was taking hold as desperate measures were employed. "Move the train cars to the edge and place them into the void", someone said. I watched as car after car were swallowed up like some fish eaten by a whale. All the cars that were available were dumped into the mine opening. We were fearful that the very hill we were standing on would also be swallowed up soon. So back to the mine elevator we went to get the latest on events. I overheard a survivor say that Willy went back for his tools and that was the last they saw of him! We were all in a state of shock...over 30 miners were MIA maybe more the crowd speculated. More stopgap measures were made such as large amounts of hay and railroad tides were used to no avail. By then, the PA State Police were on the scene and we were told to leave and go home. My Dad that night could not believe that I was an eyewitness to the disaster, we watched the scene almost all night on TV and prayed for the victims. 24 hours later, emergency vehicles were on the scene again helping miners who found high ground and one miner using his bare hands climbed a nearby airshaft to help rescue many of his fellow miners.... some good news after hours of searching and praying for survivors. My recall is that 12 people never made it back home that day and today a PA State sign marks the location and tells the story of that day that the river flooded all the mines in Northeastern PA, curtailing an

23

industry that for almost 100 years brought a living to the new European immigrants looking for streets of gold in the USA.

Induction Day, Wilkes-Barre, PA March 9, 1959! *"Do you swear to defend the laws and government of the USA...?"* Suddenly, I was a new US Army Recruit, E1, and put on a Martz Bus Lines for a ride to Fort Dix, NJ. Several things drove me to join the Army at the time: I lost my very pretty girlfriend to a summer love and still suffered from a broken heart. Also, I flunked out of college and did not see myself suffering from some low-level job that was in the mines, as a local coal cracker. The Knox Disaster made sure of that option for all of us, forever! I already spent the hot summer of 1957 in a garment factory, without any AC, working sunup to sundown, at 0.65 cents per hour, on a large stamping machine cutting out material using metal dies for ladies' brassieres. It was amazing that I did not lose my hands on that completely unsafe operation. I had other ideas on my mind and leaving PA was one of them. I wanted adventure and was ready to fight my way to a place, where it was, I really did not know. But I did know it was not in Northeastern PA's, beloved to this day, coal region!

After taking my eight-week US Army Basic Training in Fort Dix, NJ, recycled twice due to illness by the rough March weather and training from 5am until 9pm in the cold, wet NJ outdoors. After several days recovering in the hospital from pneumonia, I repeated the worst week... the Infiltration Course and Gas Mask Training. When I complained that my gas mask had no filter, and I began choking, the Sargent in charge told me to just get another one and repeat the course, so much for complaining in this man's Army!

3. First Shot at College: Penn State Nittany Lion to US Army Recruit

In March 1959, I reported to Fort Dix, New Jersey for my Army Basic Training. March is nasty month weather wise in South Jersey. Rain, wind, sometimes near freezing temperatures, can be a hostile environment when all day and night you are outside exposed to the elements. To keep from being called for KP, kitchen police duty, I volunteered to be a fireman. My job was to spend the night making sure all the heating coal stoves were operational and to act like a fire alarm, if any danger situations were to arise. It was a new experience for me as I am a morning person and look forward to a good night's rest after a hard day at work. One Army story I like to tell my kids, and now my new grandkids, was that while I was in the US Army at Fort Dix, I was ordered to go to McGuire Air Force Base several weekends to work loading large Military Air Transportation System airplanes (MATS) with all types of military supplies for shipment to Cuba. We, yes, the good old USA, from March 1959, and I'm sure before, until summer, June 1959, actually helped Fidel's Castro with his revolution! Without the USA's help, I doubt if he could ever pull it off! Cuba's Castro Revolution, of course. He even wore the same type of Army hats that all of us in the Army wore at that time. By summertime 1959, he obviously had all the military supplies he needed to win and told the United States where to get off. Meanwhile, Batista, the former political head of Cuba, left Havana with millions of dollars, according to the many books I read, and settled in Miami, Florida, USA! Thus, became the new Little Havana section of Miami that today has grown and looks like a foreign country right here in the United States. Take the truck route today, I think Route 19, into Miami and see what I mean. It is like a Third World Nation that runs for miles and miles into the city of Miami.

So, I told my kids that their Dad helped Fidel Castro with his revolution and that is not a lie, as many weekends, Saturday and Sunday, me and my fellow US Army Recruits, spent those days filling up planes at McGuire Air Force Base for their ultimate destination, Cuba.

I met many great guys at Fort Dix. Most were from the Northeast USA…from Boston to NY where they came from; all young healthy American kids looking for adventure and some just trying to do their two years and return to a mostly a then, middle-class existence. As is always the case, the New Yorkers and Bostonians stood out. The New Yorkers with their highly-developed street smarts and the New Englanders with a high intellect and superior educational foundations. Mostly of European origin and almost all Judeo/Christian religious upbringing, but proud Americans all of them! Few were as lucky as I was to get months of electronics school after basic. They were sent to advanced infantry school and off to Germany and other cold war foreign country zones as first wave forces. At my next stations, I met guys with similar MOS (Military Occupational Specialty) skills…mostly possessing what we call today as being of a technical nature or better yet high technology orientation.

Just what is "High Technology" you might have said many times before I get into the nitty gritty of my start in the field. Let me take a shot at giving you my overview. This will help you understand my personal mindset. High Technology (HT) is a business, government and industry economic sector that deals with science and primarily involves the use of electronics theory and applications. What was first called first generation involved a class called analog devices and evolving now to what we now call a digital device world. Applications range from microelectronics (transistors, integrated circuits & large-scale chips) to microwaves (radio, TV & the Internet) to large scale systems such as: GPS, Loran, ATC, Networks, the Cloud and now worldwide communications media such as satellite TV, Netflix and even Internet providers.

Pioneers whose names end in Edison, Marconi, Bell, Watson, Sarnoff, Hewlett & Packard, Westinghouse, Collins, Varian, Martin, Boeing, McDonald & Douglas,

Hughes, Einstein, Von Braun, Sanders, Goddard, Glenn, Jobs, Gates and Ellison just to name a few off the top of my head, no disservice intended as I know I missed many! All had major roles in what was, and still is today, a creative business like no other, one was discovery, excitement and accomplishment all work together, in mankind's ever driven quest, to overcome life's difficulties and uncertainties, resulting in higher standards of living for all of us!

I got lucky to get in this business early at 18, however, and I credit working with my Dad as his electrician's apprentice since I could walk that got me great scores in the Army Aptitude Testing. My results scored quite high and I was selected to go to Army Electronics schools. This was a blessing, since most of my fellow draftees went Infantry and over to Germany for the rest of their enlistments. My new assignment was to attend Basic Electronics Training at a prestigious location, The US Army Signal Corps School at Fort Monmouth, NJ and close by Bradley Beach, NJ in the summer of 1959. The new base was a welcome gift. My family, if we had the money, would visit Atlantic City every chance each summer. And I recall, in 1956, for $3.00 entrance fee to the Pier, seeing Tony Bennett live and listening to Ricky Nelson sing from outside the arena deck on the Steel Pier...I had no additional money (another $3.00) to see him live too! And, of course, the comedian Henny Youngman show was free as was Tony's. I still remember one of Henny's famous jokes on "taking his wife next year to a new place on vacation.... the kitchen!" He kept repeated "take my wife, please take my wife!" And the audience, including my family, and me laughed and laughed!

Each morning at Fort Monmouth, I admired the Airborne guys and the West Pointer's too, running everywhere on the base, in unison. I sort of regretted not going Airborne after Basic, when asked by my close friend, who did go for the additional training. My youngest son, Robert did join on his own and did 30 years in the 101st Airborne, just like my Uncle Pat (Screaming Eagle too) who was severely wounded, shot in both knees, jumping out of an airplane in

WWII's Battle of the Bulge. Uncle Pat is still alive at his writing in his 90's and got married in Las Vegas at 82 and now lives in Texas; still smoking occasionally his vice for life, king-size Pall Mall cigarettes!

In addition, my not taking an offer to attend West Point, after Army Basic, bothered me, but since it involved re-upping for several years even after graduation seemed too long an enlistment duration for me.

Fort Monmouth Electronics Training was not easy. The close-by great NJ beaches were only a destination on selective weekends. However, I was fascinated by the new training and wanted dearly to get through it successfully. In fact, for me, I learned now the hard way "how to study". Fort Monmouth had the latest classroom technology even closed-circuit TV (they had their own station too). Many a late afternoon and evenings, I took TV taped lessons on subjects I needed to master. Reviewing lessons, over and over, until it sunk in, was my method of learning. And it worked! The US ARMY was way ahead of public education and even today, lucky for all of us that it still is!

Graduation Day at Fort Monmouth, I will always remember. A full-bird Colonel was the main speaker. He told us that we were lucky soldiers with Basic and Advanced Electronics Training that would yield us careers in an explosive new field called High Technology and Electronics! He wished us all the best in our future endeavors as most of us were going on to the next level of schooling called US Army Advanced Electronics Specialist Training. Mine was to be done in Alabama, a place called the Redstone Arsenal in Huntsville, Alabama…the heart of Dixie, the Southland!

My first trips south of the Mason Dixon line began in 1959 and believe me it was like a different planet from what I was used to! After Baltimore, on my bus trip to Huntsville, the next stop was Knoxville, TN. Everyone departed for a pit stop and I saw for the first-time signs that said, "White's Only" and another sign that said, "Colored Only"! After leaving the "White's Only" marked restroom and returning to the bus, I noticed a man in the street gutter looked like he needed assistance, so I, as a good old country boy, went to assist him as I was taught to do. Suddenly, as I went to offer my help, I was stopped, and a big tall burly guy told me to go away and mind my own business. But I said he needs help…go along boy us will take care of him…. get back on your bus! I looked over and the bus was getting ready to pull out. I ran to catch it for fear of not making my schedule arrival on time in Huntsville. I looked back and two guys were pulling the man up out of the street on to the sidewalk. I then realized this was not like Wyoming Valley in PA, where it was thought to be a valley with a heart! I never forgot that scene; my first impression of the Southland was not a good one.

Huntsville, however, dispelled my initial impression, a country town with God-fearing folks and a friendly, downhome culture…one that I came to respect and enjoy. And the weather was very mild too, after an upbringing in the freezing North! Redstone was heavily populated with civilians; the military population was not like Fort Dix…it was a much smaller force…mostly high technology troops assigned to man the many missile schools located on the highly-secured base. It was also home to over 165 former German Scientist along with their leader Dr. Werner von Braun. They all worked for the US ARMY Ballistic Missile Agency soon to become NASA George Marshall Space Flight Center or now known as MSFC.

Entering for the first time to Redstone Arsenal, to the Heart of Dixie City called Huntsville, Alabama was in fact, mind blogging. Over the hills and nearby Tennessee Mountains at night, on the old Route 11, was a glorious sight to behold! The Interstate Highway system we take for granted today was passed by the Eisenhower Administration, copied from the first superhighways Ike saw in Germany during WWII. It wasn't until the 1960's until they were actually started. On Route 11S, a two-lane road, off in the distance valley was a flood-lite, football-like field, but this large stadium was filled with rockets and missiles from the US Army & US Air

Force Military Missile magnificent weapons arsenal. Coming over Mount Sano, one's eyes were riveted on this awesome sight. WOW!!! America's power was so well displayed in one place…ARMY & AIR FORCE Missiles all of them, from the Atlas, Redstone, Titan, NIKE Ajax and Hercules, HAWK, etc. Like a Super Bowl display, it gleamed in the night as though a Walt Disney Park Production. Truly an unforgettable sight to behold…simply awesome! WOW! I said again and we all repeated WOW!! I knew right then, just going on 19 years old, and seven months of Army service, that I finally arrived at my life's chosen destination and would never look back.

Photo 3.0: Redstone's OGMS – Circa 1960 photo.

Notice the WWII German captured V1 Rocket standing tall at entrance to the base! I worked behind this building where all the radars and mobile system trailers were housed.

Ever present on base was a loud rumbling sound that seemed to occur daily…. like the sound of a thunderstorm. It was the testing of the new Saturn booster over and over all the time I was there from 1959 to 1961. Several years later we would be awed by the sight of it taking our first teams of US Astronauts to the moon from Cape Kennedy, Florida. Still, today, it is an awesome sight both at KSC and at the Space Museum in Huntsville looking straight up like it is still ready to go again to places unknown!

Going back to Huntsville, 50 years later in 2009, I was delighted to see what Alabama had become, and deservedly so, the birthplace of our wondrous USA Space Program, the envy of the World! One can spend a week visiting all the space related attractions in Huntsville, now considered a World Class tourist site and attraction!

The school sign said, **US ARMY Ordnance Guided Missile School,** OGMS for short! This was my next US ARMY Electronics School assignment after 8 weeks basic, 12 weeks Electronics school now 12 more weeks of Advanced Electronics School…almost half my enlistment courtesy Uncle Sam, just going to schools. This was later changed significantly as new technology "works in a drawer" saw troubleshooting only down to the module level…then on to places like PA's Tobyhanna Depot for extensive repair. I was trained to fix things down to the component level like a common vacuum tube, transistor or other small component. Equipment field up-time was an important consideration for this new technology. Years later, we would pay dearly for changing this training philosophy due to high downtimes on the actual battlefields.

My training ranged from tracking radars to transistors from sub-systems to large scale systems eventually specializing in the NIKE Ajax/Hercules Internal Guidance System, the brains of surface to air, guided missile system designed and built by Western Electric Company (WECO), later to be called Zeus that could even take out a flying missile, like itself, even before Patriots!

My first exposure to solid state devices (transistors/integrated circuits) came in 1960 attending the Army Ordinances School Course entitled Basic Transistors by a high-tech company called RCA (Radio Corporation of America) who's founding goes back to the early days of radio and Marconi! RCA Tech Reps taught the class to advance support personnel on the Nike Ajax/Hercules & Hawk and numerous US Army Ordnance Agency equipment. The top group of engineers (rocket scientists) stationed in Huntsville work for an Agency called the Army

Ballistic Missile Agency (ABMA) as it was then known at Redstone soon to become a civilian icon called NASA. Started first by President Dwight D. Eisenhower, in 1958 and later championed by a new Boston/Harvard educated President, named John F. Kennedy, whose goal was to reach the moon by the end of the decade in 1969? Little did I know then that I would join NASA years later, even though I had many encounters with similar hardware later working at Martin Marietta's factory in Michoud, Louisiana on a Computerized Laboratory Automation System for the plant that built the NASA Shuttle Program's External Tank (ET). Also, I recall in the late 1960s at Sanders Microwave in Nashua, NH, we were asked to bid on a microwave module for a NASA spacecraft mission. After reviewing the NASA specification and request for bid, our united conclusion was that this was an impossible job and cost-prohibitive. Never did we see so many requirements in one document. Sanders did as we recommended and no-bid the job! We were working on the new F111 airplane avionics, so were not hungry for the work. We all said too much paperwork was required, typical federal government work, and that was in 1967. One day at Goddard in 1989 or 1990, we were discussing the NASA requirements and I mentioned the Sanders 1967 no-bid reply. One of my guys said that he was involved with that hardware bid package and out of 90 companies solicited to bid the job, no one… zero responded! That's when NASA got realistic with their new requirements soon and right after used the DOD type requirements like the US Navy did. I thought that was because GSFC was filled with ex-Navy personnel with Applied Physics Lab at Johns Hopkins and the Naval Research Laboratory nearby.

After graduating from OGMS Training, I was approaching one-year completion of my two-year military active duty service. Most personnel needed another year or more to get to a final assignment. I did not have enough time to be sent to Germany where most of my fellow graduates were sent, so I was assigned to remain at the School to work until my enlistment was completed. It was called an Army Field Maintenance Team responsible for maintaining the entire missile schools' facilities and training equipment. I loved this assignment. However, my personnel folder showed that I completed several drafting courses at Penn State and my skills were needed at a place called OGMS Training Aids essentially a Drafting & Design shop for the base and a place that did graphic presentations much like PowerPoint does today but with large slides and Flip Charts.

I spent several months there working in my dress uniform with tie, doing many presentations mostly for the high-level brass commanders. Redstone was the place where all US Army & US Air Force Missile School Training took place in those days.

Months later at my request, missing the hardware experience, I was again assigned to the School as a Mobile Electronics Shop Repairman. That move gave me the first hands-on experience with instruments made by none other than Hewlett Packard. From the start, I was intrigued with the high quality of these built-in instruments in US Army mobile vans equipment racks (see photo). Also, a company called Western Electric populated the rack with specialized equipment that was mobile, stand-alone Army electronic fix-it shops.

Photo 3.1: US Army Electronics Van located in Redstone Arsenal with NATO Country student

I felt that this experience was better than getting a BSEE at Penn State! And anybody, who was anybody in the USA Space Program, was here in Huntsville. Dr. Werner Von Braun's US Army Ballistic Missile Agency (ABMA), eventually was to become NASA, was there with 165 WWII German Scientist that we captured before the Russians did and put them to work here in Huntsville. I passed by their offices on my way to work every day to start all the school's equipment for daily classes.

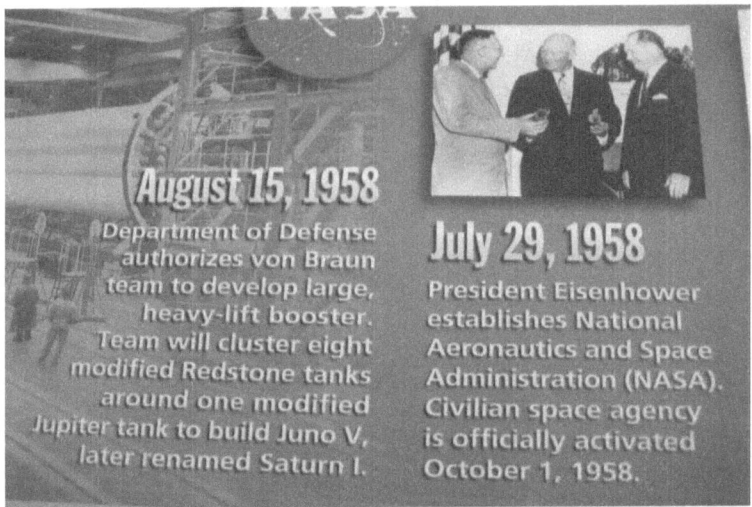

Photo 3.2: Two historical days in 1958: One for NASA, another for Saturn that would take us to the moon!

I even marched with OGMS Company B, Second Training Regiment, in the new NASA Dedication Day Parade, following then President Eisenhower's car, when he named the old ABMA now the new NASA George C. Marshall Space Flight Center (MSFC now). At work, now and every day, we could still feel the ground rumbling from rocket booster testing, down by the Tennessee River, of the future Saturn Rocket engine that eventually would get us first to the moon in 1969. My recollection at the time was that there were 3,000 civil servants and about 1,000 GIs assigned to the base.

Photo 3.3: President Eisenhower and Dr. von Braun with Col. Eifler, et al, enjoying NASA dedication.

My first Redstone location was to live in OGMS Company A's WWII Quonset huts (semi-circular prefab buildings with corrugated metal roofs) that were hot as hell in summertime with only air circulation. This was only for a brief time; as new air-conditioned quarters were quickly being built for military housing nearby. We could not wait until they were built, as summer was quickly arriving in the Deep South. Once we moved to our new housing, Army KP was ended at Redstone and contractors were hired to do those jobs. Our daily jobs were too important to miss, so this was a better tradeoff for guys like me.

Having President Eisenhower visit helped bring necessary funding for this new generation of warfare and maintaining the race with the Russians over the dominance of space. The Cold war was heating up and fast when the Russians orbited a Sputnik Satellite in 1957 before us. Dr. Von Braun and his team was given 30 days to match the Russian edge...that he did with the Redstone Rocket. Manned flight was just around the corner and the Saturn rocket was on the back burner to be first on the moon winning the ultimate race to space! Many today are critical of President Eisenhower's leadership, but he excelled in every job he held…one of our greatest heroes of the 20 Century in my book! Today, I love visiting his home and farm in Gettysburg, a real treat to any USA history buff, maintained by the US Park Service, frozen in time, as it was when he was alive.

Photo 3.4: Saturn Rocket in Huntsville at the new Space Museum

After I graduated from rocket school, designed and taught by Philco, RCA and WECO (Western Electric) Tech Rep Instructors. I earned and received a Military Occupation Specialty (MOS) entitled NIKE Ajax/Hercules Internal Guidance Repairman. However, with less than a year left in my active duty enlistment (remember that 7-year obligation came later!), I could not go overseas as other RA's (Regular Army vs. US Draftee…Me)! My fellow students could however be assigned overseas duty for the rest of their enlistment. So, I was assigned to OGMS School to maintain all the training school's electronic equipment.

I was proud to be selected to be assigned to this very impressive school. All my Army teammates were from the highest caliper levels and highly proficient in carrying out their individual duties. Close friends, however, were soon departing for foreign assignments and only seen again after our enlistments were over and we were back up north again as civilians.

Photo 3.5: NIKE Hercules Site Visitors on Tour. These were deployed at every major city in the USA in case of air attacks by the USSR!

What a good break I thought, I liked Huntsville! My new boss Army Sargent Carl Franquet, a Darby's Ranger in WWII (30 years plus guy) and even as I was to find out later when we met again in PA, He served in Viet Nam, was born and raised in my PA hometown...how that is for a coincidence! I was truly honored to work for this US Army hero! Carl recently passed away after spending his final years settling down back in PA. I discovered that he was back in PA on the Internet while visiting a Nike web-site and mentioning his name. We spent one afternoon going over old times almost 40 years later, when I was ending my career at NASA. We never forgot each other after all those years;

it is amazing how we are bonded to our true friends over all our lives. He was deservedly, ceremoniously honored and buried, with our other heroes, at the Arlington National Cemetery in Virginia.

Photo 3.6: NIKE Display – Missile and booster attached, one awesome looking sigh

4. The Cape Canaveral US Air Force Base, FL calls me...not ready for prime time yet!

The Cape was calling me in 1960...but I was not ready yet and fate had other plans for me!

In 1960, while assigned to the USA Army Ordnance Guided Missile School in Huntsville, I heard through the grapevine, that an opening was occurring with my MOS at a place called Cape Canaveral. From my teenage years, I was taken in by the great PR job of living in the sunshine state of Florida. I always wanted to visit there, but it seemed very far away, almost a three-day road trip to get there. Those teen year comic book ads showing beach place homes for $1,995 attracted my attention and I asked my Dad can we buy one Dad, please! But he looked at me like I was crazy. The new Cape job was a ground floor chance to start up future launch operations beyond just the Redstone rocket!

Photo 4.0: Many NATO countries trained on NIKE at OGMS as shown here standing on the System Radar antenna.

One of my close Army buddies also wanted the job, so we flipped a coin for the new assignment. Years later, I met my Army friend who got the KSC assignment in NYC. And he told me I got the better end of the deal. The Cape was not ready for prime time yet, he said. No air conditioning and bugs everywhere. He also had to live in an old trailer, again no AC and the food was terrible and off duty time there was nothing to do but watch the birds. Huntsville was a luxurious place compared to the Cape. My dream did come true 30 years later and the new place called KSC had all the creature comforts of home, even a new city called Titusville and with the nearby new tourist attraction called Cocoa Beach.

About one year after settling in Long Island, New York, and meeting several Army buddies from the area, I got some bad news about one of my fellow GIs who played on our Army football team. Specialist Forth Class Jerry Stewart, an Arkansas Razorback, was killed at work when hit in the head by a Hawk Missile, while it was being lifted. These were the days when hardhats were not even invented, however, soon after, wearing them became standard practice at all work locations. Today things that we take for granted are often the result of hard lessons learned some even resulting in workplace fatalities. As one often says, "only the good die young" and in Jerry's case, he was one of the not only good ones but a great friend and patriotic soldier.

Later, at Redstone, I even got to work for several months as a draftsman at the base's Training Aids Department whose job, at that time, was to prepare top brass command briefings via flip charts or just large graphics displays. PowerPoint, of course was light years away! I was a natural for doing these presentation graphics and later became an expert at NASA GSFC using Microsoft's Office PowerPoint.

I recall one special graphic being done in 1960. I prepared it for then Col. Eifel (later famous Major General, of course) the base commander who fought in the Battle of the Bulge, it had the name Martin Marietta on it with missiles flying high into the sky full of other rockets (Star Wars, I later thought?). That became known by the public years later. Little did I know I would become an Engineering Manager for Martin Marietta later in the 1980s? And, still later, spend my remaining career at NASA Goddard! Do Google: "NASA GSFC Chuck O'Boyle" to look at my last big presentation, entitled "**50 Years in Space**" done for the NASA Contractor's Annual Meeting at Goddard in 2008.

Photo 4.1: NASA Heroes and first to and on the moon.... mission accomplished successfully! *"One small step for man and one giant leap for mankind!"*...Thank You! Neil Armstrong, Michael Collins and Buzz Aldrin!

MSFC Rocket Scientist Wanted: US Federal Government career deal: 30 years and out...an offer too good to refuse! In 1961, upon receiving an honorable discharge and an early out, by two weeks, courtesy of our new President John F. Kennedy, I turned down an offer to join NASA MSFC as a GS-9. Most of my Redstone Army friends did join the new agency on the ground floor. And later they retired in their early 50s! What a great government retirement package that I didn't know about until years later, just how grand the retirement was. The Federal Government has what's called a 30 and out retirement package. I would have loved to retire in my 50s and at $50K for the rest of my life with almost full pay and medical. And then continue to work as a government contractor sometimes in the same job for another $100,000 or more! Today, in 2015, just-released government pay indicated over 17,000 receive over $200,000/yr. not counting this and $100, 000 for continuing to work until final retirement! It shows me, how someone could afford to buy those Mac-mansions, with gated communities, all around the DC beltway area! My recommendation to new college grads is to go to DC, get any government job, and in 30 years you may be able to retire quite comfortably.

One big benefit that we did receive as contractors was paid overtime at our hourly rate. So, folks made a career of working overtime even a 60-hour workweek just to get additional money. During periods of long launch waiting times and long hours of preparation, however, they earned their pay.

Photo 4.2 Hubble Space Telescope (HST) Space Shuttle Crew responsible for on-orbit repair of telescope...autographed "Chuck O'Boyle: Thanks for your support for our successful flight"! STS Crew! My Team, including me, were honored with 36 NASA Medals for this successful mission work!

Photo 4.3. Hubble Space Telescope First Servicing
Mission NASA Medallion awarded to me for On-Board
DF224 Computer Study conducted on fast track 30-day
deadline prior to launch of STS-61 Endeavor Shuttle on
December 2, 1993 to December 15, 1993.

5. Fast Forward: 30 Years to 1989 & NASA Space Flight Center, Greenbelt, MD USA

Upon coming on board at NASA Goddard (GSFC), I was introduced to my mentor Joe Bunavitch, a recently retired Gov'y (short for Government) and now working, like me, as a contractor. Joe was a complete gentleman, old school, and he became a good friend. A graduate of St. Thomas University (now University of Scranton) in Scranton, Pennsylvania. Joe like me, was a coal cracker from Pennsylvania. He started at Goddard in the beginning and worked all early projects including the moon landing in 1969. I listened to Joe, in fact, I took even took an AMA Listening Course in 1995 in downtown DC as part of my annual training. We both now worked for Unisys Corporation who had the contract for going on almost 30 years at the Goddard Spaceflight Center in Greenbelt, Maryland.

Unisys Corporation, prior to my employment was a questionable merger of Sperry and Burroughs Corporations. Imagine, a finance/banking Philly-based computer company and a highly unionized, LI, New York defense contractor being married! My best friend Frank Elia, a DEC Award-winning computer salesman and told me it was not the best of mergers! Two dissimilar products; with computers that did not talk to each other! So much for the Wall Street merger conducted by an ex-US Treasury Secretary appointed to redeem the company from some highly questionable political contributions that took place with one of them. And look at even HP today to see what sometimes million-dollar mergers take you. It seems like the only ones to benefit are CEOs who get millions of dollars in golden parachutes just to step aside or resign.

My NASA war stories started with the Challenger Disaster. As I indicated earlier, one of the reasons I came to NASA was result of the Presidential Investigation Report on Challenger Volumes I to V (science.ksc.nasa.gov). It identified many causes. One was that NASA let many Quality Assurance personnel go in order to save money.

Using my Cost of Quality model, it ended up costing NASA billions in losses, because of this flawed management tactic. Quality Assurance guru, Phil Crosby, then an expert witness and consultant appeared as a witness and said this cutback was a direct relationship to the Challenger failure. In my reading of the report, I was shocked by the admission by Morton Thiokol Engineering that advised KSC not to launch. That action could have saved the mission. Higher management at the Cape urged the Contractor to take their engineering hat off and put on your management hat on! This resulted in a go for the late morning launch. Having lived in the Orlando area for two winters in the 1980s, frost weather occurred several times often killing the many orange orchards in the area. Some farmers even packed up operations and moved to South America after multi-year losses. The Cape as it was often called then, even after being renamed for President Kennedy, suffered the same, if not worse, climate conditions. It took years for NASA to resume flights after that accident. However, after extensive investigations and upgrading of procedure and engineering changes flights resumed but with a renewal of safety emphasis that continued until the next catastrophic event.

Later, the Columbia Space Shuttle Disaster occurred in 2002! I was visiting my family in Pennsylvania when I got the call my NASA colleague Joe Osche. He asked if I had heard the news yet? No, I replied! The space shuttle was lost on reentry! No, no, not again, I replied! What happened? It seemed on reentry, after a flawless mission, something went awry, and Columbia was scattered over parts of Texas & Louisiana on its re-entry back to KSC in Florida. It took months to find out the root cause of this accident that caused the lives of several of our best astronauts (See Columbia Accident Investigation Report Volume I & II at NASA.GOV web-site.)

Going back to Challenger, according to my NASA Colleague Joe Osche, when he worked at Grumman on the shuttle cockpit design, a B1 Bomber pilot structure escape enclosure like a jet escape hatch was supposed to be included to act as protection should an accident occur, however, again according to Joe, rumors at the time were that the then Nixon Administration cut the budget and no such design was approved. To this day, he is still raising hell about that poor design! The Nixon administration I do recall also cut the budget for $250,000 to maintain the then Space Laboratory in orbit and soon fell back into Earth's atmosphere ending that successful mission as the first truly space station.

Today, 40 years later, that meager sum sounds like a mere pittance, since some 35 years later, we still give Viet Nam over $25 million a year for damage from defoiling their scorched earth from B-52 carpet bombing; surely space would have been a much better investment.

I recalled how the first time I visited NASA KSC Pad 32, in 1989, how clean and sleek both the shuttle and two side booster tanks were. And the External Tank (ET), the shuttle was mounted on was also so sleek and clean. I recall, having been inside the External Tank at Michaud, LA, on one of my many trips there, where it was made. ET was quite an engineering feat, so were the people who supported it in the manufacturing stage. Top-notch professionals, I called them after visiting several times for meetings on computerization of the operation. I recall a summer meeting, flying from Orlando and arriving in New Orleans on a 90% humidity, 90° day, when just opening the car windows, they completely fogged up! The weather there was hot as hell!

Years later, prior to the Columbia launch, I visited pad 39A again with my KSC team. I was struck by what was done to the Solid Rocket Boosters (SRBs), one on each side of the External Tank. At every circular seam, it appeared that a kind of foam was applied, very crudely, from the top of the SRB to the bottom of the entire rocket. It even looked amateurish certainly not the smooth sleek appearance I saw on earlier pad visits! The pad picture was still in my mind. What were they trying to do? Was it another what was called a workaround, I thought? Workarounds were often used as Band-Aids to maintain schedules in many manufacturing operations. I was surprised to hear this term also in use at

47

KSC. I later researched the design of the boosters and found something quite abnormal. For example, everyone is familiar with rain gutters that are on most homes in the North. They appear U-shaped for the water to flow by gravity onto downspouts away from the home. Well on the SRBs, it appeared the same application was used, however, on other military grade missiles this was designed with an inverted position along with several O-rings to seal the fit. The SRBs for cost-saving reasons, heard that one before, it was done differently. Thus, a seal was not as effective on the SRBs as it was on the DOD designed boosters. Any additional O-rings in the SRB application would be cause for concern. In worst-case weather conditions, 40° F was the minimum specifications for lunch. KSC temperature for Challenger was in the high 20s in the winter of 1986. Back to the new foam that was used to add more protection for a questionable O-ring design. Photos of the shuttle pad area showed ice sickles forming above and below the spacecraft launch area as water was thought to be a good ice melting agent. These folks, who should have known better that water use like that only worsens the freezing. Anyone born in the snow country, like me, would know that this was not the way to melt ice. However, it must be noted that KSC folks mostly were born in regions in the Southland and did not know the hazards of cold temperature or ice and water like we future snowbirds did!

After initial launch on January 16, 2003, I watched a re-run video of the launch of Columbia from a warm NASA GSFC boardroom. New stereo cams were used at the base of the shuttle for the first time. While I said what the hell is all that the debris on takeoff! One high level Govy said, it's only foam, no problem, just foam! Yes, but what I witnessed confirmed what I saw on my revisit to pad 32A earlier. Foam looking like what encircled all the SRB seams appeared to loosen up upon liftoff and spread a hail of the debris over the entire launch pad! This was called "foam shedding" by NASA Management! So much for that workaround, I said to myself! After weeks of investigation, it was discovered that the culprit behind the Columbia disaster was the new applied foam at the ET top hitting the orbiter! This debris hit the Orbiter, causing a wing hole that upon reentry resulted in dis-integration of the mother ship. Columbia was doomed from launch! One story worth mentioning was about Astronaut Sally Ride, a several times veteran of the shuttle flights, visited KSC after the accident and went right to the engineering documentation center and pulled out a book and opened it to a specific page and said, "read this sentence, **nothing** shall hit the orbiter"! And she just walked away, looking disgusted.

NASA, this time, took swift and thorough corrective action! **If it looks wrong, say something!** Posters appeared everywhere! One thing that bothered me, however, and still does to this day, was that no one person or persons were held accountable or took the blame for that accident! Oh yes, there were a few resignations, but no Club Fed time! Unlike industry, when firings happen with just a little loss or bad news, no one was held accountable for these disasters! In Senate hearings later, Sen. John McCain who I admired and respected, questioned during the investigation hearings kept repeating, I want a name, give me a name! And I said to myself Sen. consider a mirror! I told the story to the Head of Annapolis Naval Academy one night during a Navy band concert in the park on the downtown square, he replied, yes, John was quite a mover and shaker in DC politics. I did hope that he did name names, as I know a few who should have been blamed. After Columbia, word out of NASA HQ downtown, as we referred to it, was a new slogan: cheaper, better, faster was being promoted. We all commented: What the ____! It seemed that the newly appointed Administrator, coming from industry, reacted to cost overruns, and the many new missions way out of their original budgets. He wanted the escalation of cost curtailed and turned around! At the same time, the quality wave was hitting all government agencies! New quality training now became the headquarters driven practice! Everyone must attend training to understand what it was and how it could be attained! To us, assurance management professionals, this was like singing to the choir! Of course, once a bandwagon starts in DC, everyone gets on board or forever lost behind! People, especially managing supervisors, who knew little or nothing about the subject of quality, took up the torch! They became champions! It literally blew our minds; to see hated personnel become top management picks to lead the charge. We all went along for the ride; millions of dollars and thousands of man-hours were wasted with this flavor of the month government sponsored initiative! Along about the same time, new requirements were on the scene. After a new European-driven requirement called ISO-9000 appeared on the scene. At many Aerospace and high technology contractors across the whole United States, big banners appeared quote "**We are now ISO 9000 certified!**"

And now as the late and great Paul Harvey often said, the other side of that story! As an assurance professional, I too, considered this new requirement and discovered it was just a warmed-over DOD requirement called Mil- 9858, that the Europeans installed as a block to USA exports unless certified, and marketed it back to America! No kidding, we invented it, they installed it and we bought it back sounds crazy, it was crazy!

Once I get into it and what I found out later was that the many West Coast aviation industry producers used the new version and even added tighter requirements. Almost 20 to 30 more for airplane manufacturers and all of them were using it. Believe it or not, NASA a space agency would not and did not adopt it until years later! I recalled attending a high-level Agency meeting at NASA HQ in DC and talked up the benefits of AS 9000. When suddenly at the break, one of the HQ advisors came up to me and said, be quiet! We are not going to use AS9000 at NASA! So much for that recommendation! Upon returning to Goddard, I told the Director higher-ups about AS 9000 and the word back was that we were not funded to implement it. You can feel my emotions, when years later, I read in the HQ Memo that AS 9000 was now going to be adopted by NASA, effective immediately! By that time, I was deeply involved in the System Engineering side of the NASA business and

INCOSE (International Committee on Systems Engineering) was my next discovery. I was just delighted to discover this enlightened organization and how they influenced me and the rest of NASA Systems design groups.

So many books and media sources are available on NASA. Many missions, once thought of as being impossible, starting with the Mercury Seven, the Gemini Program, Apollo, Saturn even our first space hero Allen B. Shephard, USN, USA's first to space and John Glenn, USMC, who was the USA's first to orbit the earth, even up to Gene Krantz's "Failure Is Not Option" and the movie "The Right Stuff". However, little is known of the ordinary foot soldiers on the ground and what it was like at that level. Little more is known about the spacecraft world hundreds of space satellites, birds as we called them, those responsible for the day-to-day activities that are today part of our normal everyday lives. Such technologies as the international World TV, the Internet, weather forecasting, navigation, GPS mapping, communications like we never even dreamed. We today, just pick up our phone and tell our secondary or administrative assistant to take me to an Italian resort or take me to a movie or a place that we have never been before. We marvel at the little computer in our hands and all it can do for us, almost like a Dick Tracy make-believe watch we all saw in the newspaper cartoons as kids. Micro-miniaturization was invented in an American laboratory in rural New Jersey countryside, called Bell Labs! There, the greatest generation of engineers on the planet discovered semiconductors or solid-state as we know it today!

The West Coast Cape Canaveral: Vandenberg Air Force Base (VAFB) California…Welcome! Located, just north of Santa Barbara and near the Club Fed Resort at Lompoc and a few hours' drive from LAX. However, minutes by air; a small prop-driven puddle jumper, that I flew several times just to beat the traffic.

On one of my first VAFB launches, my COBE Team Leader with several assurance management personnel called me with an urgent request for my presence. It seems like the bean counters, as we called the financial people, came up with a new cost-saving technique. Instead of providing cash advances as they had in the past 30 years, they went to a check dispensing idea. People were given several low denomination checks, Visa and MasterCard, to use at in place of cash! Have you ever tried to cash a check in California or anywhere else in the United States of America? Especially, in a new town called Oxnard, California indeed what seemed as a strange community?

I recall the first time driving up from LAX to VAFB, close by the base; I passed what appeared to me as an exotic vacation resort. People were playing tennis, jogging around in athletic clothes in a scene out of Laguna Beach or Miami Beach! However, I soon found out what it was. The sign said, **"The US Federal Penitentiary at Oxnard"** or Club Fed, as we called it for short! Here the best white-collar criminals in the United States were housed. For example, Michael Milken, famous junk bond king and his pal, Ivan Boesky were both incarcerated there. And it is rumored that they made a lot of money trading on the European stock market while they were there since they were banned from trading from the US or New York Stock Exchange.

Back to my special trip, I resolved my team's problems immediately. By the way, my company suffered greatly because of the check-cashing problem. The government Team Leader eventually became a NASA Director and was one of our biggest adversaries for the rest of our contract duration. I also think she made sure that we did not win the job again after 35 years. In DC, the Fed does not just get even they have the power to also get one up! After meeting with my team and hoping to resolve this problem quickly, I come up with an idea having been a member of the NASA Federal Credit Union from day one; I asked if Vandenberg had one? Yes, of course, so I told all my team to go join it ASAP and we would pick up any charges or fees. They agreed to do it ASAP and the problem was solved! As I said the bad PR my company received from this fiasco far outweighed any cost savings that the bean counters had in mind. In my career, the bean counters always seem to be up to no good. They were always jealous of us engineers traveling all over the world and staying at hotels. They call them vacation periods! As far as the employees were concerned, in many places, the bean counters were hated and despised!

One personal recollection was while I was on the road with a former employer, a top 10 company, I was called into the chief bean counters office and asked why I rented a Lincoln for $39.95 a day from one rent the car, when I should have been rented from another for $45 a day? He continued don't you know we have we get a kickback from them rather than the one you selected? I did not know that was my reply. That the same company had fired a Vice-President for conjuring up such schemes in the past. I knew because he occupied the office next to mine for a long time. One morning I came in to work and his room was empty, a victim of bean counter mentality, I thought. The federal government does a good job with the personnel they have to uncover such schemes! Several CEOs I knew of also served time in the Club Fed for their shenanigans with government contract money. I found out that the smarter they appeared, the dumber they were, when it comes to crimes and misdemeanors. My personal values always kept me from crossing over the line and I expected all workers to do the same! I just cannot understand the mischievous behavior of managers, although the guys like Hewlett and Packard's were a rare breed of honest entrepreneurs! As some say, in the social circles of Baltimore/DC, you can never be too rich or too thin! This was rumored to be said by the Duchess of Windsor a Baltimore divorcee who took down a king!

My work ethos, looking at mission success factors that drove NASA and my management team my understanding of these is as follows:

1. Common Mindset

2. IQ Pre- & Pre-requisites for Discipline

3. Dedication & Perseverance

4. Possessing a Winner's Attitude

5. Self-managed Leadership

6. Common Purpose & Goals

7. Works for a Championship Season

8. Critical Work Skills, Education and prior Industry Experience.

A sample of my chosen Project Managers: A brief story on its own!

Photo 5.0 below shows a key part of my Project Leaders, all major NASA mission contributors! And I write this in their honor for through their expertise and leadership, I was able to be assured that NASA and my company could rely on their expertise to cover all the bases with a high degree of quality and reliability and professionalism!

Left to right: Sam Kourvaris, Gus Bonenfant, yours truly and Aaron Pokrass....over 150 total years of experience in the high technology business from individual contributor to supervisor and top levels of Management, a combination very hard to beat in any endeavor. Now let's look at their backgrounds!

Sam's previous experience long back included a brief period a Goddard, however, after finishing Johns Hopkins University with an Engineering degree, on the GI Bill, he spent most of his career at IBM raising to the Director's level. I needed a manager to run a new Project, called XTE (X-ray Timing Explorer), since I detected it contained many novice level personnel and Sam performed that job with an expertise only, he could deliver. Using a Delta Rocket, Sam & I watched it launched on 12/30/95 from the Cape and performed without a hitch! Sam retired shortly after the mission was underway and enjoyed many hours of golf at the Presidential Golf Course near DC where he resided.

Gus Bonenfant and I went back to the mid 1960 in the Boston arena. A Boston Franklin Institute graduate on the GI Bill, Gus also rose to the Managerial level at Kollsman when he worked for me while Director there. A microwave expert he fit the new TRMM (Tropical Rainfall Measuring Mission) Project after working as my Field Rep travelling all over the USA at key NASA Contractors. Launched in 1997 TRMM lasted 15 years before retiring….a tribute to the great teamwork performed on this project!

I also assigned Gus for work well done as NASA Resident at the University of California, San Diego for almost two years. I'm sure he appreciated this assignment after spending many freezing winters in his adopted State of New Hampshire. He retired to his Live free or Die state of New Hampshire prior to my stepping aside for a new team of baby boomers.

Aaron Pokrass, Philly born mathematics guru, I found from my colleague Frank Ferraro who should also be in this photo. Frank worked at the Princeton University Nuclear Laboratory along with Aaron. Both had over 30 years in the business, Aaron hit 50 prior to NASA. I needed seasoned

Manager on the Hubble Space Telescope Repair mission and even NASA blessed Aaron coming on board. You just cannot find these people any more…they just don't make them! Aaron was with General Patton's Third Army in WWII and helped liberate the many prison camps throughout Europe. A memory he seldom spoke about, but I knew it never left his thoughts of such tragic times. He worked several consecutive successful HST Repair missions as part of my staff and I and NASA was very lucky to have him on my team.

This was a brief look at just a few experts I was fortunate to hire at NASA…there was one assigned to each of the 20 to 30 active projects underway during my management reign.

Calendar Day 9/11/01 at Goddard in Greenbelt Maryland - it would be a dis-service for all the innocent victims in New York, DC and Pennsylvania, if I left out what happened at NASA Goddard on 911. First, it was a perfect weather day crisp, clear autumn day like so many choice weather days in the DC area with of course low humidity. Everything seems like a normal workday until around 9:30 in the morning, first signs of the problem came in over the Internet. Planes were off course, first from Logan then later from Dulles in the DC area. Suddenly we were startled in our engineering workplace cubicles by a reported incident in New York City at the World Trade Center. My work location that day was in a building occupied by hundreds of engineers adjacent to the main gate. Just minutes away from the many laboratories and testing activities located on-site at Goddard. I recalled a Boeing office beneath our floor and knew that a TV was in the boardroom. That's where I ran to see what was happening having been inside the World Trade Center many times, especially with my son and daughter at the famous Pub and Windows of the World exclusive restaurant located at the top floor. And remembering visiting the rooftop with my Mom and Dad, after timing the elevator to the top at 52 seconds! This complex was very familiar to me I also recall a Marriot hotel at the base and remember counting the 20 to

30 Rolls-Royce cars is parked outside the building at quitting time. Yes, it was a major New York City icon! The TV was on and displaying a flaming tower after what appeared to be a jet or missile going right through a top floor! Good grief, we all said! I also recall that over 35,000 people worked in this vicinity of New York City! Saddened for a few minutes, we said this is no accident when watching a second jumbo jet hitting the second tower. It was almost too horrible a sight to see people at the top floor who were doomed with the entire upper floors on both buildings in flames. God help us, someone in the room cried! Hopefully at least the people who occupied the lower floors could escape harm. After about 15 or 20 minutes, the first-floor tower center collapsed like a stack of dominoes. People were seen jumping out of windows some in flames! Just a horrible site for a civilized world to see him on live TV! Minutes later, the second tower collapsed with the same effect. Floors falling like dominoes, a horrific looking, and unreal scene.

Suddenly someone shouted, all had hands meeting upstairs, everybody must attend! A somber group of top managers told us to leave work and go home and take care of your families. We did not know then, at least were not told, that the Pentagon was also hit! I quickly went back to my place, packed and headed north to the PA border just in case more serious local attacks occurred. But I couldn't, as all interstate highways were closed. The only route open was to the Maryland capital Annapolis on Route 50 E! With no other escape route, I took it arriving at the capital, I noticed armed military personnel with machine guns ready at the entrance to the city. As I entered Annapolis, my gut feel was to get as far away from DC as possible while I still could. I parked my car and went over to the boat dock hoping about one was available to take me and others that Eastern shore of Maryland on the other side of the Chesapeake Bay. As I was walking briskly to catch a boat many women were sitting down eating her sandwiches on the dock. I asked where are you people from? All said, the Pentagon, it was hit too! Who did it and why were my innermost thoughts? In just 24 hours I got my answer, all who did it or executed it had their photos on every front page of every newspaper in America. We all got a large tourist boat and quickly headed east passing many police and Navy boats along the way. After several hours and with the US capital evacuated and was an empty site. On TV, we were told it was now safe to return to Annapolis. Upon back on land, I got in my car and headed north to the Poconos. There in PA, my retired parents had a home and it was a safe haven for me. Watching TV there showed what happened in New York City and in DC and over the skies of Western Pennsylvania. So many victims we did not know exact count, but in the turmoil, I dreaded the total tally of victims. NASA and DC were shut down for days, all of us try to restore normalcy in our daily lives. My World War II Dad was shocked like me and wanted to know who did this and why? We also asked, how did we have the 19 photos so quickly, less than 24 hours after the attack? Who was our new enemy? And how did they pull it off? Where was our protection?

What happened to our security? I can honestly say that I never heard of them prior to 911 attacks! Nor had any of my colleagues, this is a new experience for all of us! Worse than Pearl Harbor my Dad mentioned and with all our own planes! Such a dastardly plan clearly executed with a minimum of resources. We all thought we were dealing with a very smart new enemy that would require very smart solutions, not just answers to overcome and solve. The big cities for me had lost their lore and no longer were safe and secure as they were after winning World War II. Our world had changed that day on 9/11/01 and it would take a long time before we could feel safe again.

6. A USA Civilian Again: Now on to my 1960s-high-tech work world!

My first civilian job after duty in the US Army Ordnance Guided Missile School in Huntsville. AL, came in the high technology areas of: Long Island, New York, Plainfield & Boonton, New Jersey and Silicon Valley, Palo Alto, California. The high technology industry was just getting started. Color TV was invented; however, it was not seen in homes until 1963. And still not part of our daily world, computers were large as buildings and except for the big-league industry players at the time: IBM (Big Blue), Sperry, DOD, Burroughs, the banking business and several big DOD houses such as McDonnell Douglas, Hughes, TRW, Collins Radio, Boeing, Lockheed and others this new technology was still not a standard office tool.

In 1967, when I left HP to go to the Boston 128 area and get educated at university level, HP begged me to stay. Hey Chuck, you can be part of the new HP business computers that was just launching. Stay with us, we will make you an expert! We can send you to San Jose State and get your EE degree for just $90 for books, they added. I had already told Sanders that I accepted their generous offer and had my mind on Northeast University and getting one of those big college rings! I moved on and never looked back, even after I heard fellow workers at HP became millionaires and retired permanently in their 50s on HP stock, I loved my job in high technology, it was like an obsession to me at the time always dealing with difficult to solve problems and building organizations and using state-of-the-art tools and techniques. And as my colleagues always now say, we were the pioneers of a new industrial and scientific revolution, one that would last for hundreds of years!

Sanders was one of my best career moves! I worked for Royden Sanders twice, first in Plainview, LI, NY and later in Nashua, NH. My first time, at 21, when I was a single and unattached! Second time, as a married Father of two little boys. Here I was 27 years old, now married with two kids and going on ten-year's work experience in High

Technology! And a full-time college student at Northeastern University in Boston. In addition to, working a minimum 56-hour workweek on some of the world's newest technologies. I was a microwave technology expert at the time and one reason Sanders Microwave Division wanted to hire me. I was trained at HP's Stanford Facility; there were not many engineers that had that kind of background. HP not only taught you how to work with the latest technology, they also made you learn the theory behind the job. And the mathematics that were scientific proofs, the foundation of the technology. HP was the only place where I saw mathematical formulas on all the blackboards eventually going to white boards later. Everyone had slide rules and later HP calculators. And they were used like a carpenter uses his hammer. After attending many hours of training and almost memorizing the lesson plan and teaching techniques, I could have taught, without my formal degree, all the HP courses on the West Coast. The later place to be called the Silicon Valley! And Stanford University was the place to get your engineering degree. HP had many PhD's in EE from that great institution of learning. On the East Coast, of course Boston's MIT was where it was at! LBJ signed the Cold War GI Bill in 1967 and soon colleges all over the USA were flooded with guys like me older, wiser, all veterans of Army, Navy, Air Force or Marines.

Viet Nam, my generation can never forget that country's name! For almost 20 years of my early lifetime, the so-called police action like Korea, was to stop the far east communist world's "Domino Theory" from taking over all of Asia! LBJ did a good job of convincing all of us in the USA and world that we were right in taking all our war actions. My visit years later to the DC Viet Nam Memorial Wall made me realize the near genocide my generation almost succumbed to! Asked to go by my new friend ex-Marine Viet Nam Vet Monty, I agreed to go even though I wanted to block out all those traumatic years. I was shocked by what I saw! Not only the 58,000 named engraved in the wall, but the large Directories at each end of the Memorial containing the names and date of birth of the honored victims. DOB 1939-

1945 were the most dominant years. My generation, WWII babies most of them, not post WWII, but during the War to be more specific!

My recollection of these years and my position I will try to explain next. First, as you will read later, I was sent to Palo Alto, CA in 1964 the year of the LBJ/Goldwater Presidential Election. As a US Army Vet and not yet fully understanding the new Viet Nam War, I recall even voting for LBJ during that time. It was the last time I voted for a Democrat! The reason was that there were more than the 58,000 names of my generation on that DC WALL that were sacrificed! Thousands suffered PTSD and were hospitalized all over the USA in VA Hospitals. A near genocide of my male/female generation and to this day never truthfully acknowledged by our government. I would classify myself up to that time a hawk and conservative to the point that the then John Birch radio program from the Bay Area I listened to at lunch in California seemed to make sense! This soon changed in 1967, when in Boston, I read all the Huntington Avenue located Northeastern University Library books on Viet Nam! And I watched as students burned their draft cards and over 35,000 of draft age fleeing to nearby Canada? Could that be all wrong, I asked myself? Yes, after educating myself now as a full-time student, I too wanted to join the protestors! I became a dove in just a few months in over-educated Boston! Almost to the point of joining anti-war groups that I saw daily protesting our bombing and fighting actions. Even today, years later, we spend over $24 million a year to help Viet Nam recover from the massive bombing that we did there. Imagine this, more bombs were dropped in Viet Nam than in WWII, except for Hiroshima and Nagasaki!

Many years later, the truth finally came out! Had Jack Kennedy lived, we would never have escalated the Viet Nam War, in my opinion. Jack, as a member of Congress, visited there in the early 1950s and saw the Civil War possibility going on there and the lessons learned by the French who occupied the country for years after WWII. Who, in the 1960s would ever realize that at the turn of the Century then

US Secretary of Defense McNamara would admit on TV, that we were wrong in ever escalating the war actions! The kids, protesting on the streets of Boston when I got there in 1967, were right, and some were now dead right about the Viet Nam War!

A contractor's view of Fed Gov shutdowns is something I need to address. This occurred twice during my first 10 years at NASA. All of us contractors were on duty in addition to necessary governmental personnel. This included security and safety personnel along with critical mission operations throughout the base. And as far as assurance management personnel all over the USA remained on duty as if nothing happened. As I recall, the shutdown took two or three weeks. In that time, it was business as usual at least in the short time nothing unusual occurred. Few citizens know that civil service personnel have a union. After the so-called shutdowns, all of them received back pay for their absence; just like additional vacation pay! We contractors did not appreciate this payment, for in DC like elsewhere we held a no work, no pay position like all other working Americans!

Post-Columbia era and "the normalization of deviance!" After years of analyzing the Columbia disaster, what occurred was a change in the KSC schedule-driven culture. This was going to be a difficult change from my point of view. Coming from the DOD arena with emphasis on military specifications such as Mil-9858A and 45208 military quality practices. And with the new AS9000 West Coast Aerospace industry practices, critical checks and balances for the aviation industry, I could not agree with the headquarters and center directors for not wanting to adopt such standards and practices. Since our astronauts whose lives were at stake and always in my mindset, I used all my power to make sure we did the job right, the first time, and every time! It becomes and still is a part of my DNA and I think part of the assurance management family's culture as well.

Every nook and cranny, all NASA process stages from design, fabrication, inspection and testing to even training were scrutinized by both industry and government experts. A new organization was started and acted as an umbrella over all center activities. Mandatory requirements were re-introduced, and a new era began. Emphasis was on using world-class quality criteria an additional training and quality assurance practices and as I said before, triple redundancy in design. This practice was always utilized at NASA and now with the latest software developments, even self-correcting programs were fixing errors was built in the hardware design for added mission reliability. I myself became an expert reviewer after joining the INCOSE (International Council on Systems Engineering) organization and keeping up with the new developments in both hardware and software practices. One example I worked on and will share with you was on the new NASA JWST (James Web Space Telescope) Observatory. I, after some digging, found out that the software was being bought from a Boston 128 high technology house that had so much debt on their balance sheet that I didn't think they would survive until launch time. It seems that I was not the only one concerned, I found out that the DOD also was using this company software and on new tactical weapon systems also. Soon after voicing my concern to the new project director, it was announced in the monthly space newspaper that Big Blue (IBM) had acquired the company. All the new company founders became rich overnight! So much for the power inside the beltway. As they would often say. I went on to review the new project software development plan and tore it to pieces. All standard software standards were lacking from the beginning. I wrote up my report and asked for a complete rewrite to fix the holes I discovered. Time and time again, once I got back as an individual contributor role, I was appalled how a once mission success culture had deteriorated so quickly! The cheaper, better, faster false mantra had taken hold and the structure was crumbling before my very eyes. What was needed was to quickly return to the "right stuff era", back to the basics! Easy to say, but the players from the era were

63

now for far and few between, like myself, were all approaching the retirement years. It was, as they say, a hell of a ride and a quite a fabulous journey while it lasted!

I noticed today a new culture focused on their PCs and handheld devices, oblivious to what was happening all around them. Maybe that earlier generation boss had a point about "computers were for secretaries" for we never lost focus of the building of the hardware in those early days of NASA. 50 years had passed and hardening of the arteries was taking place, the patient needed rejuvenation. Today in 2015, nearly 8 years later, I hear that the focus now is on Muslim Outreach! As General Patton said after World War II, I'm glad I'm not part of it the 21st Century work world.

Yet remaining in place and still standing is the organization we put together, mirroring the government, in our industry type organization structure. It's hard to beat something that we worked for almost 25 years. The last of the old team members are now in charge of carrying on with the same mindset but straining to hold on to proven practices. Today's culture was initially charged to break it (the system) and then fix it! Whereas mine was if it isn't broken, don't fix it! And today's quest to level the so-called smokestacks is doing just that, even the old management and organization practices are being thrown out identified as being obsolete.

Tracking NASA Satellite on-orbit anomalies for a 10-year timeframe was a new job that I tackled. As the new Mission Operations Assurance, we recorded tens of thousands of spacecraft systems anomalies. These anomalies occurred on all NASA spacecraft and tracking system events. Our job was to log all anomalies by spacecraft and report them weekly to NASA management. Critical events were reported immediately or if non-critical reported in weekly reports by specific on orbit spacecraft. NASA has almost 50 active spacecraft such as HST, GOES, and TDRSS etc. to addressed, all providing active performance history. And like human being's regular health checks, all NASA's spacecraft have their same health checks. We were to look for recurring patterns that after engineering analysis provide future corrective and preventative measures.

High technology tools are extensively used in the NASA workplaces. From the very beginnings of flight, the business had handwritten drafting and design tools and techniques. Eventually, they were upgraded to computer-aided design (CAD) in one fell swoop. As I said earlier, as a seasoned manager and computer user since the late 1960s, I was shocked with not being able to use the power of my computer in my daily tasks as I entered NASA. As an MBA, I was very lucky since both HP and DEC computers were used in all my courses at Babson College. My Class of 1973 was the first to use computers in every course we were required to take, from matrix algebra to management information systems, all required and contained a computer application.

Babson at the time, and still is today, a number one ranked school in the world for its emphasis on entrepreneurship. In my earlier management jobs, I helped apply computers in purchasing, inventory control, quality assurance in all factory operations. And I even started a computer program facility at a local university in New Hampshire. Today, they are even offering PhD courses on the subject.

Once our civil servant boss, who was anti-computer retired, the floodgates opened. I was fortunate to rapidly catch-up by taking every computer training the Federal Government had to offer. Soon, I too was using Microsoft Office Excel, PowerPoint and Project Manager in all my work assignments. And, once I discovered Visio, it too became a powerful communication tool.

Figure 6.0: U S Army Hawk Missile Display with other US Arsenal Missiles (Redstone, et al) in background at the Space Museum in Huntsville, AL

Back to the Huntsville and what went on there. The Raytheon Boston area made Hawk missile was the newest addition to the OGMS School Training. Hawk was a predecessor of the now famous Patriot missile system, so valuable to our allies overseas since deployed years ago. Highly mobile, and multiple battery capability it is the best reflection of American technology of the 20th Century. Many of my close friends were Hawk trained and most of us Yankees somehow gravitated together. From Boston natives to New Yorkers all of us Northeasterners became friends for much of our lifetimes. A common culture and common values played a big part of whom we bonded with. Plus a few southern friends fit right in with us on all our social activities. A few of my northern born colleagues even married local girls and stayed at Redstone their entire careers raising to high ranks among the NASA MSFC government personnel.

7. Bye-Bye Huntsville! Hello PA, LI, New York, New Jersey & CA!

So here I am in 1961, not yet 21 years old, after getting an early out by the new President Kennedy (to save money spent later on the Bay of Pigs debacle), with years of aerospace training/experience under my belt, but of no value to the coal region of my beloved Pennsylvania. After, a few weeks of receiving state unemployment checks, seen by many factory workers there like a paid vacation, I just got bored. Nothing to do but drink Rolling Rock and BS with my many unmarried, now Vet friends (since they did not have jobs either)! However, all of us did beat the draft and we thought we put that obligation behind us.

The new beloved President Kennedy, my Dad's generation and a War II hero, had his hands full with the Russians and the Cuban Missile crises. I thought surely the Army would call me back then, but it waited a few more years after the Viet Nam TET Offensive! I was married and had two boys in 1967, when I got my notice to report for active duty military service!

During preceding years, I left PA and quickly got a job in Long Island, NY, a then hotbed for technology jobs. The company called Sanders Associates, Inc. (great name!) was founded by one of those Steve Jobs/Bill Gates college dropouts of the 1950/60s. Royden Sanders became the head of Raytheon without a degree during WWII and taken out of college for the job! He was one of my heroes, leaving Raytheon with several colleagues to form Sanders Associates, Incorporated. His insight brought today's video games, a billion-dollar industry today, but also computers into the workplace. It all started when he asked his engineering department to give him a view into that large computer in the basement of his new corporate tower...desk top computing was invented here too with his new division called Sanders Data Systems. Computerized reservations systems such as the Wizard of Avis, ECM Systems for Naval aviation and microwave miniaturization were all part of the wide array of technologies developed at Sanders.

Eventually, I worked in all those areas, but it all started, for me, in 1961 in Sanders Plainview LI, NY facility. My job as Standards Laboratory Group Leader, started by checking all the new HP test equipment bought for the US Navy. I got to know the HP test equipment and HP Representatives quite well especially, when I questioned their HP Quality!

In late 1961, November to be exact, I almost "bought the farm"! Spending a Friday night on the town with a new driver, a childhood friend who had been drinking and was driving. And I was in the passenger's seat when he hit a telephone pole on the way back to my car that was parked in town. No seatbelts were required in those days and I hit the windshield and almost bled to death waiting for a Doctor to arrive at my hometown hospital to stop the bleeding and patch me up. After spending two years in the Army without getting a scratch this was what resulted from just partying with my friends? I was given last rites and my parents were quite upset with events from my weekend off from new job in New York. After a month convalescing, I returned to my job with facial scars that can still be seen today on my forehead. I did settle down right after that event and found what I thought was the right girl and married shortly after in the spring of 1963 to my bosses Secretary from Sanders Plainview, LI, NY.

In 1963, I was living in Plainfield, NJ and working at the Lockheed plant on Route 22. For a short time, we were off-site at a location in Clarke, NJ. My Father coincidently worked behind my building at Clarke Roller Bearing Division of GM as an electrician. He was still commuting weekends to PA and had a rented room in as private home in Westfield. Two events were soon to occur that would change both the World and the state forever. On November 22, 1963, I was at the Lockheed Laboratory when word was spread that the President Kennedy had been shot in Dallas! Kennedy was like a rock star, loved by the masses, but as I was to find out later despised be some underworld figures and a nobody called Oswald an ex-Marine, who was married to a Russian born citizen. Well for several days over the long weekend we were glued to our new little 19" Philco black/white portable TV! What we witnessed live was history in the making that later would become the subject of movies and books for decades, even into the 21st Century! The assassination was soon covered in detail and Walter Cronkite's brief statement became history that "at 1:00PM at the Dallas Memorial Hospital, President Kennedy was dead?" Next day on live TV, as we were watching to see this guy Oswald, another shooting by Jack Ruby local club owner.... Oswald was also dead! What followed was a full week of mourning; culminating in the funeral wake at the Capitol Rotunda then burial march to Arlington Cemetery and lighting of the Eternal Torch. We kept those days and week's events permanently in our minds for the rest of our lives for the Camelot Days in America were forever changed and the age of innocence was over for America!

The second event occurring just a few months later and as I was watching TV from my new home in Mountain View, CA. This event was the nation's civil rights riots that occurred in every major city in the USA. I watched on CA TV as martial law was declared in New Jersey and mobs of rioters were trampling the very Apartment grounds in Plainfield, NJ that we just lived in being trashed. And thanking God for our being thousands of miles away from the chaos of the East coast cities that were now a dangerous place to be even near. Months later, we returned to the Mountain Lakes/Dover part of the state as a now HP employee, after things calmed down and living conditions returned to near normal in the Garden State. This was not to last too long as the Granite State of NH and Boston area was soon to be our new home for almost the next 20 years!

Today, as I told my Babson College grad school friend, fellow MBA teaching colleague, Professor Bob Tropea, ex-State Representative from the Granite State, that NH needs to erect a big statue spot-lighted 24/7/365 of Mr. Royden Sanders and place it at the Massachusetts/New Hampshire border entering Nashua! He did more to help NH recover from the old mill economy than anyone in the state and brought a whole new generation of engineering talent to the great, Granite State.

From NY to NJ in one job-hop!

After two more years at Sanders Plainview, LI, NY and many raises, every few months, I was still single and getting tired of going back and forth to PA on weekends. Most PA natives did this for years. In fact, PA has more retirees who were born there returning to their home state than any other state! My Father for example, commuted daily too far off places of work because my Mother hated to leave PA and who can blame her! It sure beat Jersey and NYC for safe lifestyles and family ties, not to mention the great ethnic foods available, especially Italian! In my commute to NY, I always passed by Lockheed Electronics on Route 22 in Plainfield, NJ. Formerly called Stavid Engineering, it is today, a large shopping mall on the famous or infamous double deuce as they referred to Route 22. Years later, it merged and moved to Sanders Nashua when Lockheed bought Sanders for $1 billion, a bargain deal, as far as I was concerned. Later, it too was sold by Lockheed (for $3B) and now is called BAE Systems. What appealed to me was the two hours closer to home in PA…a big benefit for me to go home to PA! After leaving Sanders, Plainview, LI, NY in 1963, I worked at Lockheed Plainfield, NJ in a Laboratory calibrating exotic equipment mostly in the microware sector. I also attended many HP Training seminars on microwave instrumentation that was becoming my specialty. I liked Lockheed; a west coast managed organization of professionals that gave me several raises in one year after joining. Plus, after leaving the company the month before, they still paid for my Son's big bill at Muhlenberg Hospital in Plainfield having been born prematurely and remaining in the hospital for the first 30 days of his life.

My son Chuck Junior, I call him the Lockheed baby, because they paid for his entire first 30 days of life in the hospital, today he is a New York bar member with the Michigan Law School degree. After I moved to New Hampshire in the late 1960s my family became heavily involved in what was

70

called the Pony Club. All members had horses and of course socially mixed with the wealthy neighbors of my new bedroom community of greater Nashua. My son Junior, at the ninth grade entered nearby Lawrence Academy and did quite well academically. So much so that he was selected at Amherst College after visiting Yale, Princeton and Haverford. I never could take advantage of the Ivy League institutions since even as a teenager, going into 9th grade, even getting into even Penn Military Academy and Scranton Prep schools both prestigious, prep schools in PA, in 1953, was out of the question. I applied and when the school representative visited our house and said to my father can you afford $2000 a year, he almost choked! I think he was only making $3,000 or $4,000 a year at the time, so much for my early education and thank God later in my life for the G.I. Bill.

My younger son, Robert a chip off the old block, is a story all by himself. I'm quite proud of his many accomplishments! Another manager in the family, he worked his way up the ladder, without my assistance, to a managerial position in a world-renowned Nashville, Tennessee automobile company. I did help him out briefly, by providing a good high school education and paying for him to finish Bishop Guertin High School in Nashua, NH. Formerly, and all boys high school, and now coed, it gave him a strong foundation to successfully complete his MBA later from Western Kentucky University, unfunded by me. At 18-year-old, and from a broken home caused by a war of the roses divorce, Robert managed to make it to my parents' home in Pennsylvania right after his high school graduation. My Dad, bless him, took Robert under his wing. Robert wanted to stay in Pennsylvania and enroll again in high school to play football a game he like his Dad also loved to play. He was, searching like us all, just to live a normal life and hoped to find it with his devoted and loving grandparents. Next thing I know, living in Florida at the time, was Robert was soon to be in the Army thanks to my Dad efforts again. Robert went on to spend 30 years in the Army eventually as a Drill Instructor for the Screaming

71

Eagles, the 101st Airborne out of Fort Campbell, Kentucky. His marriage was blessed with four beautiful girls and one boy. Robert was devoted to those kids even after his divorce after 15 years of marriage.

My last offspring, to the best of my knowledge, was a new baby girl that I just adored, born in my senior year of college. We named her Erin, a name that reflected her genes and the grand Irish race! After graduating from an upscale New Hampshire bedroom community high school, she enrolled, with her horse, at a special private girls' school in Virginia called Sweet Briar College. I often tell my close friends that my daughter and her horse both have BA degrees with honors! She went on to be a world traveler like her Dad teaching computers to Lawyers all over the planet…first class I may say! Retiring at 32 years old, to her adopted state of Texas with a new family, I told her that her Mother, a native New Yorker, beat her when she retired at 20 years old and married me!

My devoted Sister Peg, my only sibling, also spent most of her life working for a High Technology business too! At RCA, Semi-conductor for almost 40 years, soon to be spun off and now after a few mergers is called Fairchild Semi-conductor out of Portland, Maine. As an Administrative Assistant, originally a Secretary, having attended a local Secretarial School, it was her first and only job. I still tease her today, since even my generation was used as almost slaves to these large Corporations who earned millions over the years and gave peanuts to their devoted work force. Once you were locked in, it was difficult for you to find another job, especially in the coal country of Pennsylvania. She never received a pension, since remaining on a fixed salary worker all her life working there. Lucky for her that the 401K plan was instituted early enough for her to save for retirement for on a widow's SS Pension, it is very hard just to survive. I still can hear her pleasant voice answering the lobby and main phone for years, "Welcome to RCA, how may I help you!" That is now replaced by a computerized

message! Press one for English; Press two for Spanish, etc.!
If you know your number press 3!

Now a Florida resident, she enjoys her time in the sun and
talking on her new iPad to her Grand Daughters in cold
Pittsburg. As she tells me often, she doesn't miss PA at all
and loves her beach home and the Florida heat all year long.
And I say come on Peg, I know you miss the pizza and all
the good folks of PA and Penn State football too, even
though it will never be the same after Joe Pa's death! You
can certainly leave PA, but PA never leaves you, is why they
all come back there to die, for no one knows them in Florida!

8. California…here I come! And Back to Where I started from!

One day talking to the Hewlett Packard (HP) Representative he asked me if he could call me at home. I gave him my home number and at 6 o'clock sharp the phone rang. Chuck, how would you like to join HP he said! Wow! Of course! I replied almost automatically…ok he said when could you start? Well afterward, I found out that in those days HP asked people to come to work for them, this was their practice. I was elated and replied, "how about in two weeks!" Ok, we want to send you to Palo Alto for training and will arrange everything, not to worry, HP will take good care of everything including your family who will go too! And they did as promised; HP was the only company I actually loved in my career. All who I worked with at HP treated me like family. Bill and Dave had the right approach…no wonder they grew to be bigger than IBM over the years. Today, I am sad for what the foreign banks and incompetent management did to that icon of Silicon Valley…Bill and Dave would roll over in their graves if they could see it now!

I was 24 at the time I moved to California in 1964. Inside the sun filled plant (with saw tooth roofs to see the sky) I toured the entire facility. What a work environment, green before there was a green revolution. And in a spotless machine shop, a big banner hanging high near the roofline that said, **"Vote Goldwater '64, No bread '65 and No water '66! Bury Goldwater"!** HP was not shy in letting personnel do their thing politically…a sign of the times back then!

I spent months in HP Customer Service under a boss named John Walling, retired Navy Officer, who I knew from all the HP Seminars I attended. At the time, I didn't know that I was being groomed for a higher-level job, since I was a loyal hard-working HP employee and a new generation on the HP Team. I made many friends in Palo Alto and HP became my new family of work smarter, not harder social network.

One day having lunch with Cliff Jones, one of the HP Supervisors, who wrote several technology training books, a real down to earth guy, gave me some important insight on the California high technology culture. He knew how awed I was to be at HP and said something that I remembered my entire career. Cliff said to paraphrase him, "that these people were all human like us, they put on their pants every day one leg at a time and are lucky to be at HP where they can do their thing working in a new technology that gave them fame and in the future fortunes too"! He was right on; I never thought I could not compete in any job that I would be assigned in the future…at HP or elsewhere! They were all like Cliff at HP, smart, highly educated, but still humble folks at the center of the Silicon Valley Universe, before it became what it is today, like the rest of the USA.

Each Friday, I was invited to all the best places to dine free! All we needed to do was show up, a HP tradition in those early days. And all places were the best, we were part of the HP family and the company appreciated all of us. One weekend I wanted to visit Yosemite National Park, but my old 1956 Buick I bought for $275, was not in the best of shape. I sold it months later for $650 dollars! My boss told me to go down to the HP motor pool and get a brand-new Chevy, fill it up on us and bring it back when you return! This was HP in the 1960s. HP head count was under 2000 and they just hit a $200 Million a year sales level. Compare that today with over $100B sales and several hundreds of thousands of employees worldwide. As I said many times, now even bigger than IBM!

HP spent almost a year in California making sure I could do my job as a microwave instrument product line Senior Engineer. I learned to rebuild instruments like a car assembly line from scratch, from the simplest to the most complex…even microwave spectrum analyzers, my favorite one at the time. I loved my job, my company and my new life! California was the place to be in the 1960s…a lifestyle like nowhere else on the planet!

California people are the best talented, best educated, and most likeable that I have ever met. Most of them came from the east coast and harsh climate states like me! A hard to beat work world culture all on its own and I was part of it…most of the lessons I learned stayed with me like a toolbox to be used when I needed them. Bill Hewlett and David Packard were my Father's Greatest Generation people…all gone now, but never forgotten. I recall year's later reading that Dave came out of retirement at 80 to rescue his baby for the highly educated managerial succession plans (BSEE/MBAS) did not work out…and he turned it around in just a few months. Dave was the most dynamic person I ever met, and I did meet many along the way. His company was seen at the time, the very best company to work for as most of my colleagues retired in their 50s with the millions made in HP stock over the years. If you bought 3 shares the company gave you one, free! And you also got profit sharing checks every Christmas many in the thousands of dollars. One look at the HP parking lots would tell you HP was the best places to work or play as many people did…but not on the job! With my first profit sharing check, I paid cash for my first new car (1965 Chevy Impala, stick shift) and we worked harder and smarter to make the next one better.

One motivating meeting, I recall worth mentioning was a town hall discussion with the top man himself, David Packard. He came into the room with sleeves rolled up (not a good sign, as that was a signal that he meant business). He started out by raising hell about borrowing from the banks to grow the business. He gave the salesmen a charge…no more banks! I want you people to increase your sales by 20% next year so we can fund our growth from within, was his message! Yes, they listened! The following year sales were up by 25%...I guess they all got Dave's message!

After attaining a HP Senior Engineering title with no degree, I felt something was lacking with my education. My fellow colleagues all had degrees from the best schools in the country. Remember when Steve Jobs tried to get into HP after a summer job, they told him to go to Stanford and get a

BSEE and they would think about it! Steve got even and became bigger than HP, in fact, the biggest in the world today. Perhaps if HP hired him, he would have lacked the motivation to do his own thing as most of us in the business were in those days. We all had families and lucky to be where we were…employed with the World's best company, HP!

It was 1967, LBJ passed the cold War GI College Bill for military veterans, the least he could do for Viet Nam Era victims. I always wanted to finish college specially to wear one of those big rings engineers at Sanders Associates wore with *Northeastern University Founded 1898* on it!

Photo 8.0: Class Ring 1971 NU, Boston!

Go East Young Man….go East!

By 1967, I found myself back on the East Coast in Boonton, NJ! HP bought Boonton Radio Corporation and moved the Eastern Regional Service Department into the new building. My training was over now. I and about 100 other Palo Alto folks were assigned jobs there to cover the Eastern USA. Note: After the first Jersey winter, I was the only one left, all others returned to California after a very cold winter and much sickness. They wondered how we ever survived in such a nasty climate!!!

After my Palo Alto on the job training assignment, I was to report to the then new Eastern Regional Service Center located in Boonton, New Jersey. All of HP's east coast work was sent here for repair and recalibration. HP had the resources there to completely refurbish all the HP product lines and my contribution was with the microwave offerings. My job involved RF coax and waveguide products and later I repaired all kinds of signal generators and even became an expert in backward wave oscillators and the most sophisticated top of the line, Spectrum Analyzers became my favorite!

My first week on the new job was for me an awakening! I then learned a life-long lesson that to this day is a powerful influence…I learned "The HP Way" to do work! I completed my first instrument repair job on an old HP product that needed upgrading. And passed it along to the final station required before returning it to the buyer company. When it came time for my job to be checked, I was called down to have my work reviewed. At that time with several years under my belt, I thought I was a hot shot and felt I did a good job on my work assignments. This was not the case however, as a HP Final Inspection Leader called Larry, wrote up many "oversights" or faults on the instrument that I worked on. His list was long and together we reviewed what he found. I later, found out that he was a former Lockheed aircraft inspector, an expert who was sent East to make sure things were done here "The HP Way"! Well, after that embarrassing situation, my learning curve error plot went way down. I looked at each finding and started making up my own checklist of things and fixed everything that Larry wrote up. And in the future, I would self-inspect all my work, the HP Way. before saying it was finished. In fact, I made it a goal to have Larry find zero defects or errors in all my future work results. Later, Larry had a hell of a time finding any oversights on any of my work. Larry and I later became the best of friends as we were very much alike in our demand for high quality work results. I owe a gratitude of thanks for him instilling in me and my work ethic, "the HP Way" of doing things.

LBJ signed the GI Bill after much pomp and circumstance. In my estimation, it was the only good thing he ever did! I even voted for him, the last time I ever did go Democrat!

Ten years of Viet Nam convinced me he needed to retire. I went to Boston, in 1967, a war hawk but after reading all the Northeastern University books on Viet Nam, I became a dove. Boston youths protesting the war were right on! When I saw former Sec. of Defense Robert McNamara on TV in 2003, apologize for his role and hearing the Viet Nam General ask him "we never wanted help from the Chinese...we fought them for centuries don't you people read history books", I knew I held the correct view point. But it was too late for my generations 57,000 names on the Memorial Wall in DC.

In those days, I got so many technical magazines every month and read every one, cover to cover. I also digested all the HP Technical papers and General Radio Journals I could get my hands and collected all of them. One thing that I knew was that this was a fast-moving business and I knew to survive, I needed to keep on top of everything new in my chosen career. Electronics News, then a Fairchild publication, was as important a read as was the NY Times and Wall Street Journal. When teaching, I even got a free copy along with Barron's Weekly.

Microwave Magazine had a new ad **"Come to New Hampshire and Work for Us!" Sanders Associates Incorporated!"** Well, I was over 10,000 people who signed up to work at Sanders from those advertisements. This move is just what I was looking for. Boston was one hour away and with the new GI Bill...I was headed to Northeastern University!!! With my previous Sanders experience and now HP training and experience they made me an offer I couldn't refuse. HP tried their best to convince me to stay.... we are going to be in the computer business.... you will be on the ground floor, but my word was final.

9 – The New Hampshire (NH) USA Granite State "Live free or Die"!

I moved to New Hampshire two weeks later to the Sanders Microwave Division on Simon Street in Nashua. That same year, I had a custom-built house designed and built a short distance from work so I could come home at lunch like I did in Palo Alto at HP. Not too bad for being 27 years old. I was seldom at home however, going to school full-time and working 56-hour workweeks and doing my homework. I transferred all my previous courses from Fairleigh Dickinson, Penn State and NYU Farmingdale and entered Northeastern as a sophomore. My aspiration was to get an associate degree; just two years and that would be that! I got it in 1971 and continued for a BS, since I was paid to go by the Army and collected also from my employer too! It was like having two jobs at once! The ***"Success of Failure"*** book was the title of a best seller in those days, I had been there myself, and now was making the Dean's List every semester and graduating college with honors. I worked hard to be the best in every class and it finally paid off big time!

Photo 9.0: Boston Campus of Northeastern University – GO, Huntington Avenue Huskies!

While assigned to Sanders Microwave Division, I oversaw all test instrumentation. This was a natural job for me after spending years at HP Customer Service. Mr. Sanders, the then CEO, acquired a UK firm who made microwave test instrumentation hoping to compete with HP at least in the European marketplace. MESL was in Edenborough, Scotland. I was given one of their first "production line" instruments and using my HP training began to tear apart the somewhat amateurish design. With several pages of design and build shortcomings, I submitted them to the General Manager. Next thing I know is that I am on an International flight to London's Heathrow Airport followed by a puddle jumper flight to Edenborough, Scotland! Sanders made all the flight preparations including my passport, flight tickets and hotel/car reservations.

It was 1967 and the Boston Red Sox was going for the pennant. Also, I got to see my first James Bond movie in London in a small down-home movie theater. We arrived on the weekend and was able to do London town. While touring, on that weekend, I walked around Trafalgar Square and wanted to see where the new Playboy Club was located. Just curious, since I did not have a key for entry or membership. Suddenly, as I approached a new street made of cobble stone, I experienced something I never did before. I did not know what it was until years later, preparing for a graduate school lecture on Creative Decision-making. I came upon a phenomenon called "de ja vu"!

The cobblestone path or ancient road where I was walking on was built by the Roman Legions who occupied London at the time. I recall stopping and told my colleague, an older Senior Engineer who reported to another division, that it seemed like I had been there before! Yes, I continued, around the corner I envisioned an Apothecary shop! Let's go see, he replied! We continued walking and what do you know, it was an Apothecary Shop just as I pictured, unknown to me before and only my mind's eye! I experienced what they call de ja vu but at the time, just could not explain it. The ground I was standing on was built by the Roman

Legions during the period of the Roman Empire and in my mind, I had been to this very place before. My foot signaled my memory and walking the cobblestone-built road triggered this probably in my body's DNA was my explanation. I returned to London several times later, however never experienced that unexplainable event that first occurred in 1967. Oh yes, we later discovered the Playboy Club Casino in a tall conservative looking building but were disappointed we had no key to enter. Years later, in the Windy city, I did breakfast in Chicago at the original Club downtown by then ready to close due to lack of business.

On to Scotland on Monday and the charming city of Edenborough. The University sits high on a hill overlooking the city and, on the day, we arrived was the Feast of St. George! My hotel was not a hotel, it was a large castle outside the city limits going back to Medieval times probably where Mr. Sanders and his entourage stayed when visiting here for their golfing trips. Some say that he purchased MESL just to golf nearby. I say that if he did more power to him, he certainly deserved and earned it!

My stay was quite memorable, The President and VP of MESL were my hosts and were delightful chaps to say the least! My purpose was to show them what I learned at HP as far as design, manufacturing and servicing their products. And they listened and implemented all I showed them.

On my trip back from London to Boston, I boarded the plane and looked into the cockpit (as I always did look for some grey hair) and our chief pilot looked like a WWI German flyer with monocle in one eye and a shaved head! He got on the intercom once airborne and said "Welcome to Flight 67 destined for New York City! Everyone suddenly got out their tickets to check the destination! Hell, I'm on the wrong plane I said to myself! Minutes later, The Captain corrected himself...pardon me he said...I mean Boston!!!Wow! We were all relieved! Once close to the USA, the pilot put on the Red Sox game over the intercom just to redeem his error for his Red Sox passengers. We arrived safely at Logan Airport, all glad to be back home in the good old USA again.

By 1972, one year after getting my Associates Degree, going summers, I got my BS w/Honors! Now I still had time left on the GI Bill. I heard so much about an MBA Program at Babson College in Wellesley and graduating with honors had no problem getting in. We are honored and delighted to have you come to Babson, the Admissions Director said when I interviewed for placement. Like HP, I felt like Babson was part of my family…a very prestigious college whose claim to fame was that the founder saved the Kennedy's from the 1929 Crash…Joe picked a winner with Roger Babson as his financial advisor. In 15 consecutive months, I graduated with Distinction, a new MBA with emphasis on Computers and Entrepreneurship. It would not be the end of my academia years at 33 years old.

College was not over for me, not quite yet! I spent the next ten years teaching mostly graduate students (MBAs) in at least 10 courses that I designed ranging from Computers to Management to Finance, Decision-making and even Economics, including Reaganomics! It was an honor for me to do this work and I treated it as a hobby while also working full time at a Director's level for high technology companies in the greater Boston 128 corridor.

Photo 9.1: F111 joint force (US Navy/Air Force) Tactical Jet Fighter in flight!

The Sanders (SA) experience for me was invaluable…from Plainview to Nashua; I got involved with all electronics from solid-state components (microelectronics) to Microwave from Computers to Electronics Countermeasures Systems (ECMs). In addition, the entire manufacturing pipeline from the plant receiving dock to the final shipping dock and all stages in between. My short-term goal was to become an Engineering Manager. SA Microwave Division was quite a place when I arrived. The new miniature, microwave modules for the F-III Joint Air Force/ Navy jet tactical swept wing fighter, were in production. Large antennas S, X & Ku Band were also in production. The satellite dish business was yet to evolve. The test equipment was state of the art. No hand recordings of results were allowed. All data needed to be printed via a hard copy for Government review. I was amazed how advanced Sanders Associates were in all areas of product testing. In fact, my old HP boss toured the factory and he too was amazed on what advanced methods that were used in a production line basis. Such were factories all over the Boston Route 128 area…there were over 200 colleges and Universities around the 128 area that yielded many, many top-flight High Technology personnel.

I met one of my best friends Professor Bob Tropea at SA Microwave; we also worked at SA Data Systems, commuted to Babson, Wellesley, MA and eventually even taught college together in the same MBA Program. In fact, in midterms of 1983, I relocated to Baltimore and he was asked by our good friend and former Babson Professor, Dr. George Shagory, now head of our MBA Program, to take over my classes. Bob still teases me about those times and my sudden bolting out of my new-found home in NH and heading south!

Photo 9.2: Rivier University, Nashua, NH Malloy Hall - The new 1970s MBA Program helped rescue this college, now a prestigious University!

I designed my MBA courses so I could apply Harvard Business Review reprints as training aides. I made many trips to the Cambridge based Harvard Coop to search for my class material. HBR reprint titles of the time that were very popular among the students were:

> o How to Motivate Employees
>
> o One More Time, How to Motivate Employees
>
> o Leadership Styles

Also, in my Financial Management Courses I & II, I had subscriptions to the Wall Street Journal and Barron's as reading materials. And since I had been exposed to the Harvard case studies at Babson, I also use them in my classes too. My job as Lecturer, was to make my courses both interesting and informative applying theory and practical business world practices together, in a synergistic manner.

My MIS Course had several textbooks, one containing many large-scale computer operation case studies on business applications to give students a real-world look at how technology was taking inroads in the business world.

These were some of the courses that I designed, developed and taught at the graduate level over almost 10 years at Rivier College, now University.

1. **Computers in Business** - Having taken Electronic Data Processing (EDP) I, II & III a full year's course at Northeastern University and Linear Programming & Matrix Algebra using computers at Babson along with 9 other computer-based courses, I had no problem designing the course for first time users. And, in addition having a computer in my office at Sanders during my entire graduate school training where I used the Dartmouth College Timesharing System (DTSS) for all my research and studies. I emulated Babson's Computer Facility and was asked and chosen to start a similar one at Rivier which was highly sought after. We started using the DTSS and installed several terminals in a small closet-like classroom. Far from what you see today a complete Computer Center with all the bells and whistles. Sister Theresa, a math major graduate, of my first students, did quite a job after she took all my courses and made the Computer Center one of the best in Southern NH!

2. **Creative Decision Making** – a mind-blowing adventure for all students was a new course never offered in NH to the best of my knowledge. Babson had one of the best Professors teaching it if you can call it that...I better describe it as coaching it for most of the lessons are non-traditional and innovative in nature. Few of my students ever acknowledged training that focused on hot to think and be creative! In fact, I asked several hundred over the years if anyone ever had taken a course on how to think and only once did, I get a reply, he was a West Point graduate at the US Army Military Academy! So much

for USA educational institutions.... no one left behind...everybody now graduates from college! It's now become for some students, the high school of the 1950s!

3. **Financial Management I** – My course at Northeastern was very boring. Although the Professor (who worked for the insurance industry) did his best with the little material they had at the time, I wanted a different approach to be used to make Finance interesting and exciting which it really is once you study it. Using classic annual reports from the Fortune 500 and the Wall Street Journal, we covered current events that were making headlines as well as theory and basic application of financial tools to analyze financial statements. Accounting I & II were prerequisites for enrolling in this course, so basic debits and credits were not included in my offering. One new term called, Zero-Based Budgeting, was covered and even today needs to be practiced in both the public and private sectors of our USA economy.

4. **Financial Management II** - was a classic case study course that covered both private and   public ventures and how to analyze and apply financial management tools and techniques to both assess and plan business ventures. Cash Flow Analysis was covered, and cash management exercises demonstrated its importance to an enterprise. Both short and long-term views were considered, and recommended courses of action resulted from analyzing individual case scenarios.

5. **Organization Behavior** – One of my favorite courses to teach. It involved what drives us, what motivates us and gives us answers that often are first discovered from studying this critical subject for all managers to take. Using Harvard Business Review Classics supplemented by the latest textbook on the subject this course shows

why college level training is important for any leader to take.

6. **Management Information Systems** - This course required taking computer fundamentals as a prerequisite. Emphasis was placed on Large Scale Computer Operations understanding using the systems approach and applying it to business enterprises and public institutions. Case studies of actual real-world applications were analyzed, evaluated for effectiveness.

7. **Macroeconomics for Managers** - Supply side economics and implications were studied. Reaganomics in vogue at the time was reviewed and analyzed. Consumer + Investment + Government issues and present and past-day practices were reviewed and debated. Federal Reserve Banks actions were monitored and their contributions to the US Market economy were followed.

8. **Marketing Management** – Product, place, price and promotional elements are covered with emphasis placed on optimizing the mix during business cycle periods involving both controllable and uncontrollable factors that are constantly changing in the US & World business marketplaces.

9. **Operations Management** – A multi-disciplined course covering purchasing, production control, inventory control, quality assurance and entire manufacturing applications using both case studies and specialized textbook learning tools.

In between my college classes at several institutions in the Boston/NH area, I often would go out of town, together with my then close friend and former Babson Professor George Shagory. By then he was Director of the MBA Program at Rivier College! We would do training seminars at Babson, Vermont and NH State Police Academies and Northwestern University in Evanston, IL. George did the Management

Basics and I followed him by delving into Budgeting and Revenue Generation Exercises. We always were welcomed back to do more training of leaders at all these places. George also started a Management Institute at Rivier and was the first to do "Women in Business" Seminars that I also participated in doing my mini-course offerings. He was a top-level Business School Professor, honorable Veteran and Naval Officer, and after his passing, his family created a George Shagory Scholarship Fund at Rivier in his honor.

Early in my career at Sanders Microwave Division, I worked for a General Manager, who had a PHD EE from MIT in Boston and had a large rubber stamp on his office desk with the words **"I know it can't be done"!** He used it often to get lower level manager's attention. Whoa be tied, if it appeared on your in basket. For the result could mean early termination of your services. He even hired PHDs and called them idiots in front of all of us just to make his day. And he was big enough to get away with it! We all got even however when he was fired and was escorted off the premises for using company resources for his personal benefit. Something like reverse casting eagles and owls and selling them for profit!!! It seemed to me the meaner and more ruthless a GM was in those days the better the Board of Directors like him!! It was a rare occasion that I met any more professional and nice people like I did at HP during my years there. And I recall on Route 128, hearing the name, Jack the knife at GE, as his victims sometimes called him, who did not get that reputation from being a nice guy!

President Harry S. Truman had a **"Buck Stops Here!"** sign not bad for an untrained college leader. Harry never went finished college and made incredible decisions for all of us in the USA. And President Ronald Reagan's famous desk saying read **"I know, it can be done!"** I even bought one of those after visiting his Library in Simi Valley, CA. I liked them all and, in my career always sided with Reagan's. Looking back and hindsight is 20/20 someone said and here's where I think cultural upbringing and cultural values make a big difference in one's career

At this point you must be thinking why didn't you break away and do your own thing? You had the education, background and experience. Oh yes, but I did three times! The first, of course, was my own, a new venture called Small Business Management Assurance (SBMA). It came from a thesis paper I was assigned to do to complete my graduate studies. At the time, I completed my MBA at Babson College in December 1973, the business and government world buzz was about a new retirement package proposed by the Federal Government in DC. It was called the Individual Retirement Act or IRA for short. Eventually, ERISA, the Employment Retirement Income Security Act of 1975 was passed. The IRA was a plan were individual taxpayers could avoid taxes by putting away from each payday a sum of money for their retirement years. After retirement, they could withdraw funds at a lower tax bracket and have a nice nest egg to complement Social Security. Later 401(k) s came next along with the 403 B's. These are great plans for every dollar saved annually companies would match a certain percent too. To me it was a golden opportunity to get in on the ground floor specializing in small business organizations that make up the biggest piece of total business pie in the United States of America. Today, over $7 Trillion are in these accounts and IRA rollovers are over $300 Billion!

Yes, for many years this Fed Gov 401K looked like just the fix for those faced with later retirement. But, as with all recent government plans, long-term implications and large gaps that the Wall Street crowd had in place such as guard bands and downside protection in case of a falling market were serious lacking. When the business cycles turned down millions were hit with big losses...1987, 1998, 2000, 2008 were just some years of distress. My family relatives lost hundreds of thousands of savings hit by big drops in equities while still paying for professional assistance in picking the so-called winners. As for myself, I did get hit once, embarrassingly, as an MBA! MY eye was off the ball, in 2000, for just a few days while I was working almost around the clock.... a minor hit some may say around $30,000! I, to say the least, was very upset. I promised myself, this would never happen again! And to this day, knock on wood, it hasn't. All the folks who were hit were mostly uninformed, hard-working people; ignorant of the many forces in the stock market and certainly at the mercy of what they thought of as being out of their control. Had they had sell prices like the pros do, all their losses would have been protected. However, it was like the slaughtering of the lambs' every time the market went down. $401K plans were in my estimation the biggest scam of the 20th Century. Soon to be followed by the next biggest one called Affordable Health Care in the 21st Century. Pretending to help folks with their retirement and luring them to put hard earned dollars into a high-risk situation with dreams of achieving a fortune was just another Wall Street/DC K-street scam in my opinion. And forcing a US citizen, under penalty of fines, to spend hard earned dollars for plans that have such high deductibles that they are useless and costing a mortgage payment to avoid getting fined! So much for big government run organizations of our times. Both are zero sum games for all us worker bees and suffered by millions of uninformed citizens!

My new MBA Business Plan vision, in 1973, was to focus on small entrepreneurs, who needed help implementing these new retirement plans. The plan was to start regionally in the greater Boston area then branch out to all New England. Eventually, I could cover the entire United States. My Plan Thesis received an "A" from my Entrepreneurship Course. It was never returned to me! In hindsight, like the Steve Jobs and Bill Gates did, I should have rolled the dice! Eventually, the big players entered the market, people like Fidelity Boston, Franklin Templeton, Edward Jones, and Vanguard who now are world leaders in retirement planning for major USA corporations. As they say first come first served! My energies went into the sure revenue generator, at the time, fast ROI, college teaching which I did for 10 years on the side as my second job.

Once I began my side job college teaching, my good friend Bob Tropea, soon joined our newly launched MBA Program.

It grew like a firestorm. Three groups dominated enrollment at the time: Digital Equipment Corporation (DEC), Wang Labs Corporation and the Regional FAA Center located in Nashua. We were the first MBA Program in the greater Southern New Hampshire area. And had excellent cadre of professionals who were very active in the business community. The Program went from double-digit enrollment to triple digit populations in a matter of three years. Now, a big-league MBA Program Director was being sought out and what do you know, our Babson Corporate Strategy, a capstone course, Professor got the job! Dr. George Shagory became a close friend and relocated to a newly built home in nearby Nashua.

One day, all three of us, Bob, George and I sat down and started a new business called Stage IV (Bob's catch name) Corporation! I became Treasurer, George the President, Bob VP of Operations. We focused on business consulting solutions. New Hampshire was becoming Mini-Silicon Valley and we wanted to be part of the new high technology business sector. *For any business to be successful, several critical factors are required*: **First** a viable, vision of your offering and the unique business idea are of vital importance! **Second**, time, something a rich man cannot buy, but the poor man has plenty of, as the saying goes! And **Third**, of course, is $$$money... start-up capital as the venture capital guys say and for that you need a comprehensive detailed 5-year written business plan.

Stage IV Corp. had only one or perhaps two of these critical factors. Our time was limited, both Bob and I, had full-time manager jobs in the Aerospace/DOD industry. Dr. George had his hands full with getting the new MBA Program off the ground. Plus, each of us were doing seminars and training at other colleges and training academies all over the United States. Mine were in White River Junction, Vermont, Concord, NH, Evanston, Illinois at Northwestern University, and Babson even at Rivier College too. Stage IV was a good business idea, if we had and devoted our time and energy.

Yes, like the garden, all hard work is necessary if anything was to grow and blossom.

Later, the last venture was attempted in Orlando, Florida, where not just millions of dollars were being made, but billions! A new venture was began called Neutech (since defunct), however, at the last minute, I chickened out. And lucky for me I did, as we shall see later! Neutech, Inc. began with a marketing expert Monty, who had a full-time job as a Marketing Manager; Frank, a digital engineer, who was Salesman of the Year for Digital (DEC) and yours truly, now a Systems Engineering Manager for a new business startup at Martin in a town called Maitland, a bedroom community just outside of Orlando, with a grand view over Lake Destiny. All of us were already working together for a new business during the day and seeking our own after hours, hopefully, to make the big bucks someday like many others were in the new high technology center of the USA Southland!

One story that sort of puts Orlando business community perspective is as follows. Last year, 2014 for Valentine's Day, I took my wife to Orlando to see some of the sites and Circa de Sole' in particular. We arrived at Pop Disney where I stayed years earlier for $50 a night. I picked the 1950s section. Pop Disney is a 20th Century, decade-focused, hotel complex centered on the 1950s, 1960s and 1970s. These sections are frozen in time, so one can see what it was like to live during that decade. Great idea and new business venture. I went up to the front desk and asked for a room. Do you have a reservation I was asked? No, but a few years ago, I stayed here when it was new. Do you have any rooms? I asked. Yes, sir, we have 35,000 rooms at this complex, one of several at the park. But we are booked solid, perhaps we can reserve you one in the future. No, I need one tonight! By the way, I replied what are the rates for a senior citizen? Without a blink, and with a straight face, she said they start at $175 a night. What? I said to myself. Here is where my MBA degree kicked in. Let's see $175 x 35,000 equals $6,125,000 x 365 equals $2,235,625,000. Now, I could see how their CEO gets $50 million a year as the former one did for almost 20 years…. yes, $1 billion salary as my Apple iPhone said after multiplying it. I drove out of the complex and said let's go home back to Tampa, my wife cheerfully agreed being native Chinese…. for her too, 1,138 Yuan one a night for a room is just insane! But my sister's son as well as millions of other tourists, thought nothing of spending $1000 a day taking his two kids' wife there for a week with his company's bonus check! But, remember, we all think nothing for filling up our automobile gas tanks for ~$70 or $80! When I was in high school, $0.65 in my tank lasted five days to and from school at $0.11 a gallon and besides you got free green stamps and a free gift plus cleaning your windshield and checking your oil and all your tires, all for free! Oh, how I loved and miss those 1950s!

Neutech was founded by two my friends, Monty and Frank with me as a silent partner. Monty could sell igloos to the Eskimos and was a marketing manager for many years before he threw himself on the sword one day at a computer show in Atlanta; unfortunately for him with me not by his side. Frank was Electrical Engineer for DEC Computers and was recently awarded a trip to Hawaii as Salesman of the Year selling Martin Orlando over 25 DEC computers in one year. Frank and I were like brothers over the years and still contact each other over the web. Today, my wife Julie and I visit Frank, in Orlando, several times a year bringing my homemade Italian delicacies. After a recent back operation, Frank would never be the same again. Once a dancing with the stars-like champion, it takes a lot out of me to see him this way and recalling him years ago, on the dance floor enjoying his bachelor life.

Frank had a great sales and marketing background and worked for Ken Olsen's Digital Equipment Corporation (DEC) CEO of the Year even on Fortune magazine's cover. One of Ken's downfall, however, was he could never see a use for a home computer, it just wasn't a profitable market. Dell, Compaq and HP sure made a liar out of Ken who was not one of the new generation people like Gates or Jobs.

Ken, however, was an excellent CEO for other reasons. His DEC Company Tech World in Boston was one of the biggest events held annually with so many people they needed to bring in cruise lines of boats to handle the visitors. Just like an Apple was back in the day.

Monty, Frank and I had a good vision of the future high-tech world as clearly shown by Bill Gates' DVD video on the future. At the time, Monty's wife had other ideas…such things as flowers, household decorations, etc. that red flagged me to just back out! Thank God, I did! After several months, Neutech had its first crisis, the IRS! It seems that the small business they acquired had liens on it and they were not found by their incompetent attorney at the business closing deal. Frank and Monty were into paying a $40,000 tax bills or else. That killed Neutech and Monty returned to the industry market to keep his head above water. Oh, I forgot the Atlanta incident causing his demise of marketing manager came because he called his boss an AH out loud at a computer show. He was immediately fired and never really recovered. I think he was even blackballed throughout his career. Last time, I found out that he was selling fire alarms for home use in the Las Vegas area. Such a tragic ending for a highly talented salesman.

Next story on the journey: Nashua, NH and Sanders Data Systems (SA) – Desk Top Computer Pioneer –Wizard of Avis Reservation System terminals producer!

Data Systems was a new spinoff of Sanders Associates. It focused on large-scale reservations systems and several US State law enforcement agencies data base development and recent deployment. Such products as the SA620 and SA720 Data Display Systems were preludes to the desk top Personal Computers of today. And they were all made in USA, right in the new South Nashua Facility that housed it and the Corporate Headquarters Tower as we called it. Product application-wise, at the time, I recall going up to South Portland, Maine (Great town!) and visiting a large insurance company that housed hundreds of female workers typing on old-style mechanical type-writers. This was the market that

95

was ripe for computerization and Sanders had the next generation products that could replace all that I saw.

My job as a Sr. Quality Engineer was assigned to work on the developing the first digital key to tape SA System 6000 Data Recording System. The prototype, almost completed, was sent to a Canadian insurance company for evaluation...and got shot down. By this time, I had a problem with the new management team after my old friend and boss Korean War decorated Vet Ed Roberts passed away an early age (very heavy smoker, heart attack victim). I wanted to get back into the Military end of Sanders business. The Canadians had a basic problem with the System in that it could not play back to the system TV type screen...however; it was called a data recording system. I was only assigned to the new project in the last few weeks before it shipped and when I told the Contractor head engineer, replied to my question of why no playback? He answered, if you told me that a few weeks ago we could have fixed it. I found out that consultant guys like that stretched the job if they could and received extremely high pay for what they did. Four years later, I found out he joined DEC as a VP of Engineering. And one of my friends told me to go see his new mansion, custom-built in the town over from me. We were struck by the opulence; stone-built with a pool in a very expensive new up-scale complex built for the hundreds of DEC people leaving Massachusetts to beat the corporate taxes...the locals don't call it Taxachusetts for nothing! So many moved that the conservative NH state went from strict Republican to now Democrat, something I never would have predicted this would happen to NH, never!

My Plan B: Go to work at Sanders Associates (SA) on Canal Street, Nashua, NH: Defense Systems Division

My new assignment was to become an expert in semi-conductor technology as my new boss told me. I want you to read every Mil Spec on these devices and help find the best manufacturing house in the USA for us to use on our new Electronic Countermeasures Systems (ECM). I did just that and soon was called on to answer all the engineering

96

questions or find answers for them by using the mil specs or seeking out industry experts on such matters.

Remember, micro-miniaturization in electronics circuitry was pushed to the maximum by the DOD and NASA! For spaceflights payload size and weight were critical design constraints and factors. In addition, reliability was another critical design parameter and an ever-demanding end product feature and goal. As I mentioned here my Sanders Associates ECM story on how the old point contact detectors failed prior to the new Schottky barrier diode. Detection then was made by whisker-like diodes that were leapfrogged in Mean Time Before Failure (MTBF) by using new HP solid-state devices! Solid-state was far more reliable than the original vacuum tube designs where heat, power and high failure rates when exposed to hot/cold temperatures and harsh environmental in what we called shake, rattle and roll applications. 20th Century first generation designs went from using vacuum tubes, later solid-state diodes and transistors, followed by integrated circuits (ICs) then large-scale integration (LSI) and now complete digital circuit functional building blocks that make up products from hand-held phones to high-definition TV and even today's great Apple iPad and watches!

My first big job getting back into the defense business was to accompany a Team of Sanders Engineers to tour the USA microelectronics industry plants to determine where the NAVY could buy parts from for their new fighter Jet ECM Systems. After doing a two-week circuit, travelling from Boston area suppliers to Bloomington, Indiana (home of Purdue University) to Dallas, Texas (home of Collins Radio) and then to Silicon Valley producers and finally to the LA area all homes of the final selection list. I picked three companies who deserved to be highlighted as the best of the best, although we visited many more in our quest to find the very best military suppliers.

USA Winning contractor, my Number One pick! <u>Collins Radio</u>, Richardson Texas, now called Rockwell International.

One of the top high-technology companies in the world, Collins Radio was another company I needed to visit for assessment purposes to become a Navy ECM system parts supplier. This is a quick-reaction contract brought on by the Viet Nam War's enemy TET Offensive Campaign. The Navy wanted none other than the best parts to build systems to avoid SAM D missiles and taking down our carrier deployed Jet war fighters. Without countermeasures, our brave naval aviators were sitting ducks flying over regions were SAM defenses where deployed. One can only imagine being threatened by a Patriot missile today, formally a HAWK-type missile. And flying over areas protected by this viral threat. An analogy would be like duck hunters on the Chesapeake among hundreds of hunters armed by laser-guided guns with shotgun type payloads exploding in thousands of lethal pieces.

I was to first go to Collins Corporate Tower building to get my badge and visit Security. It was a 30-story building. I was taken by the empty parking lot and almost empty building. The feeling was very eerie, and I was told by one of the security guard to take the elevator to the top floor. Every floor up had all offices empty of employees. All desk and office equipment were still in place, it was right out of the Twilight Zone TV show, an empty city abandoned! Some 300 executives occupied this Corporate Mahogany Row Suite of high-powered offices. All were fired weeks earlier and now they were on the street.

Upon the elevator reaching the top floor penthouse offices obviously, the former home of their CEO, I was met by the Rockwall executive in charge and his bodyguard! "Welcome to Rockwell International", he said, "please come in it"! He was a short man and his bodyguard towered over both him and me. I was a potential DOD customer, bringing in new business and was treated professionally during my entire stay!

Old man Collins must have been quite a genius. Especially after seeing his in-house computer facilities that impressed even me, who had seen the latest state-of-the-art computer operations all over the USA. I counted 50 large machines networked, before networking was in vogue, around the world! A 24/7/365 operation moneymaker that transacted business globally just like it was downtown. Amazing, was my innermost thoughts!

In a courtyard seen in the rear of the tower is where Collins built his own lunar lander spacecraft. And invited NASA to come visit and told them is this what you need for the moon landing years before Grumman Aircraft made the real one. The place was an engineering facility and manufacturing operation that made everything from microelectronic parts to large-scale avionic and aerospace systems. Collins was a national treasure. Where is he, I asked? Oh, we let him go is the answer. What! These people must be crazy! I thought to myself! Collins was thrown out of his treasure trove of amazing high technology.

I became very sad, imagine you're a Bill Hewlett or David Packard or Thomas Watson being thrown out on the street by takeover corporations, a definite downside of American business practices. After all, years later, even Steven Jobs was thrown out of his baby called Apple only to return many years later and make it bigger and the best in the world

Rockwell kept microelectronics operation intact. Nothing was disturbed and I saw why. It was a world-class facility probably one of the best in the World! They were a key provider for supplying military grade semiconductor and microelectronic parts on the planet. I remained sad during my entire visit. For I could envision this company at its best time, running like a clock, but now just a shadow of itself. But still reflecting what it had been, a great place to work! I rated Collins now Rockwell among the top three out of all US facilities we visited. They were Super Bowl level players in microelectronic engineering and manufacturing.

I only write about the top three companies I picked from all the places we visited on our two-week tour of USA semi-conductor houses.

<u>Teledyne Microelectronics</u> Los Angeles, CA a sister company that I would later work for almost 5 years as operations manager, had the best production line operation of all top players. Every station was automated and produced very high volumes product with the very high yields of 98% or greater. This was the key to profitability in the microelectronic business, at the time; yield or what is a reasonable percentage of products that makes it out the door, after undergoing stringent environmental testing. Hot and cold thermal cycling followed by testing, shake, rattle and roll testing, loose particle detection testing, all are required prior to shipment. High yields in the 90s were a very profitable level to be attained consistently. Teledyne had fabrication and environmental testing down pat. At the time, most of the work they were producing was for DOD applications that added to manufacturing complexity.

Another Number One house! <u>Howard Hughes'</u> <u>Microelectronics,</u> Newport Beach, California – Hughes' operation that I had the pleasure of visiting or for this matter, auditing, was very impressive to say the least! Just when you think you have seen the best operation then comes Hughes! Visiting later in 1989, at Long Beach, they were celebrating the 1939 World's Fair exhibits and among the sites were the Hughes-built Spruce Goose aircraft. I could not pass an opportunity to look inside this magnificent wonder. And I was awed by what I saw. This was a 747-size aircraft large enough to hold hundreds of soldiers inside a whale-like body…and it flew close to here one fine day to prove its worth. The Howard Hughes organization in California was an American original. The organization chart shown to me illustrated how Howard Hughes was at the top of the Hughes Medical Group followed down too many organizations ranging from microelectronics to aviation all sectors of technology were covered. Each block showed a stand-alone operation in everything they needed to make a high technology product. My focus was on their microelectronic division presented by a charming, charismatic German-born and educated General Manager, who was an expert in his chosen engineering field. Only a few times in my career that

I would meet managers that I would love to work for. He was one of the best! HP was the other one, however, Hughes was up there too! It had a magnetic quality as a place where the winners were, and technology was tops!

Photo 9.3: NASA Lunar Lander designed and built by Grumman Aviation, Bethpage, LI, NY

More facts from Sanders Federal Systems PRIDE PROGRAM ALQ-126 Electronic Countermeasures System – Canal Street Facility

In 1970, I needed to settle on the title and role of Reliability Manager. Most of my staff of engineers had MSEEs from schools such as MIT and did not appreciate me being their new boss, like Royden Sanders without a degree! But Sanders, was in deep trouble and needed a test-engineering expert to oversee what was revenue "stopper" situation. One of the big fixes David Packard did as US Deputy Secretary of Defense was hold the contractors' feet to the fire. He introduced a "Fly Before Buy" policy, whereby DOD Contractors needed to prove their designs prior to shipping before government acceptance was signed off. It seems the first unit tested by Sanders failed this first test. Using

101

measures such as demonstrated mean time to failure, the system fell short of meeting that specification which was 160 hours.

The first thing that happened was my staff added to and was gifted to include two new Rensselaer Polytechnic Institute new PhD EEs, with the latest computer training under their belt. I immediately put them to work off- site working on doing worse-case design analysis of critical operational functions. In addition, one of the highest failure modes involved old style whisker type detection diodes that opened when exposed to temperature cycling. My connections at HP helped me connect with the new semiconductor operation called HP Labs. HP was working on solid state replacement diodes, but the yield was less than expected for commercial use. I asked if they could supply a few for our application. However, unlike the few dollars part, these were hundreds of dollars more expensive. But US Naval Aviator's lives were at stake. And we got our top management approval to buy a few. By the way, I was asked to report to a retired Associate who was providing oversight to other Sanders Associates. His name was Jim Levan. He was a fine, older gentleman and we got along just fine…I valued his support and he helped me to cut through all the red tape and political games played in every organization.

Once we replaced the old design parts with solid state, we were over the hump! System reliability was attained; however, we were not done fixing the failed systems. After just a few weeks, my PhDs hit a home run too! They found and was the case with all products transitioning over to solid state circuitry, that circuit flow diode protection was lacking, and several paths were like fuses causing failures at every path usually caused by turn-on/turn-off electrical spikes that surged through the circuit paths. I recalled a similar problem at HP when the new generation of Stanford EEs had the same problems with the first generation of solid-state products. Just when we completed our upgrades, they were signed off by the Executive VP at Sanders Harold Pope…soon to be company President. I needed to phone in progress to Harold daily until we finished our work. He would call me back if he found any errors...and that was only once where I made a minor one number addition mistake. He was one of the best managers at Sanders and a true gentleman. My college undergrad days were ending around the same time. And believe me when I tell you, I was hated by the program management team who took a back seat until we did our engineering critique and analysis. I recall even upon leaving Sanders for the last time to do my thing on Route 128, walking down the big mill halls when fellow managers did not recognize me or even thanking me for saving their jobs...jealous and envious of my accomplishments to the very end.

Even though I still did not have my college completed, my experience at HP and other places helped achieve a few months turnaround...and Sanders was back on track shipping ECM Systems directly to Viet Nam Navy Carriers for installation on naval aircraft. These Systems were lifesavers for pilots; who I heard would not fly unless they had Sanders ECM gear on board.

10. Boston 128 Arena – Turnaround Management 101 & Fast Tracker MBA!

After completing my MBA in 1½ years going two summers, I was itching to get to the next level of management and run a full department on my own. Although I could have remained at Sanders for my entire career as a Reliability Manager, on its hottest product at the height of the Vietnam War, with electronic counter measures systems (ECM). After all, I did have the program support team responsible for proving to the Navy, in house, that our products worked beyond their specifications. And I did help rescue a failing technology desperately sought by our US Navy flyers, carrier jet pilots, home all the then, Top Guns!

Years later at an Aerospace Industry Association of America (AIAA) Convention, I bumped into my former Sanders boss who by now was a Director. He was a cool dude who gave me my first Managers job at Canal Street and rode my hard-earned victories for all they were worth! He later went on to Babson and got his MBA too! He also told me, one of my new employees I hired, a Rensselaer University PhD. in EE, had made the Director level too! These were two of the good guys; I had the pleasure of working with back in the day.

The Boston Sunday Globe each Sunday was then in the early 1970s, a smorgasbord of career opportunities. Looking for what I wanted, and I found it on the first search, Quality and Reliability Manager wanted by Infoton Corporation in Burlington, Massachusetts.

I knew where it was located right off Route 128, near my Northeastern University campus in Burlington. I thought my commuting days to 128 were over after my school years, however, that was not to happen for several more years, five more years.

I aced the interview with the VP and General Manager and all Directors in the hiring process. "We need you!" they all said, and the top man asked, "When can you start?" Accepting the first salary offer is often a mistake the movers and shakers in the business say. However, the jump was substantial not even to include the benefits in stock and bonus dividends if we exceeded our monthly production quotas.

Our main product was a standalone, desktop, first-generation computerized, cashier-type desktop terminal very much like today's desktop PC. Sears, nationwide, were replacing all their then mechanical machines with these faster computerized devices. I had charge of the production testing activities along with all the Q & R functions. The department consisted of incoming inspection and in-process inspection & test, final systems test, and quality/reliability engineering were all part of my organization. One thing that I did not miss at NASA was this industry end of the month crunch to get the highest quantity and quality of your products out the door before the calendar turned over. When organizing my function, I designed a production control chart showing production flow and station status at each stage of manufacturing. The General Manager took one look at the chart and adopted it immediately! We used that chart daily in our management meetings. The GM was quite complementary to me for doing that chart, as it was not part of my departmental responsibilities or my duties. I found out throughout my career, that going the extra mile whether in sports or business always paid off and, in this case, it did big time!

Sears's chain of stores in the USA was our big customer. They all were being modernized and their stores were upgraded with our new computer technology. My micro-electronic experience at Sanders also paid off big time! I started a semiconductor burn-in program at Infoton, with all incoming chip material. The first batches had abnormal failure rates of 30 to 40%! Small companies like we were at the time, got delivery of the bottom of the barrel when it came to electronic components reliability. Some were even rejected lots returned by other companies. Unless the company invested in burn-in programs like I initiated, they would eventually fail, like the first digital watch companies succumbed to. For every two watches they produced, one was for warranty claims. Eventually you go out of business with the situation

In addition, I visited our suppliers and told them "no more junk!" If we receive it, we would be sending it back with no payment made, and I have the President of the company's support in this action. The company was a classic case study for high technology need for a Quality Assurance organization. Profits suffered, shipments were delayed, and due to manufacturing rejections and failures and product returns from the field were very high. I watched how, with just a few basic fixes and improvements, resulted in self-correction of the production process. Our previous out of control operation was fixed after about one year. The factory was almost optimized. Also, using just-in-time and MRP inventory, we also helped increase cash flow and are shipments were for the first time, on time, month after month. Again, a classic Harvard case study on how to turn around high-technology operations and I was now part of the solution to getting there.

Management is not a 9 to 5 workday! More like 7am to 7 pm was my workday for years. Not counting teaching days, when I arrived at home at 10 or 11PM! One day, when I came to work my boss was a no-show! He was always there before me and left after me. As GM, this was his routine, and it happened the next day again… MIA! Well, is he sick or what is going on, was on everyone's mind? On the third day he was back and invited me into his office. Chuck, how would I you like to be my new Operations Manager? We often discussed my career aspirations at lunchtime together, he knew that my next function that I wanted to manage was plant operations consisting of manufacturing and production and all the associated activities such as purchasing, inventory control, production control, quality assurance and even field service.

He continued, "I'm leaving soon to become President of Teledyne's 128 operations". I'd like you to come with me as my Operations guy! How can I not say okay! He almost doubled my salary and Teledyne stock the best performer on the New York Stock Exchange for 10 years running.

In two weeks, I was in the Teledyne Corporation family as Operations Manager! My work problems more than doubled. We produced machines for the entire semiconductor industry from chip pick and place to semiconductor assembly even test stations. A whole new world of automated micro-miniature manufacturing machines that were to be workhorses of this fast-moving semiconductor industry. After I started plotting weekly bookings, new orders, it was almost 18 months before the trend started to show an industry turnaround! It was a boom or bust type of market. Almost like in the old GM driven economy days. Whatever GM did, the rest of the country did! If GM was in trouble the entire USA economy suffered! The same analogy occurred here when microelectronics was still in their infancy stage. This was before the PCs and the high definition TV era was still in the embryo stages of their life cycle. However, the best was yet to come. We dealt with this drastic downturn in bookings by doing several smart marketing and management decision-making.

First, we stopped buying material, took stock of our material, increased initiatives for early customer's payments and computerized our factory inventory. In addition, we slow rolled payments to vendors and went on a shortened workweek for production line personnel. Marketing-wise, we instituted a new mini-line of economical products that smaller worldwide producers could use. I also recommended a sales contest that was immediately adopted. In addition, the big company saver, a bold move, was to do a Pareto Analysis of our inventory and sales of our products spare parts and selectively increase margins on these sales items. Big ticket items for our product lines spare parts came to our rescue! This one action saved the company in the deep decline in orders and it was my boss's big idea. His experience was why he got paid the big bucks and drove a new company supplied automobile every year. In the end, we were honored as one of the most profitable of the Teledyne family a year later, after recovery occurred and our economic measures were still in place.

Back to my adopted home, the Granite State! State Motto…Live Free or Die…no state taxes, sales or income, but one of the nation's first lotteries and the best placed highway liquor super stores on the planet!!

It was now 1978, and my commuting back and forth to New Hampshire all was getting to me. In addition, teaching was now not only one or two courses per week to four to five, even Saturday morning. Also, I only I did not need to make this two-hour hectic Route 128 commute, I would be able to enjoy my new place in the country and all my now growing family and their new activities.

A Long Island New York company called Kollsman Instruments, planning to relocate to southern New Hampshire was advertising for new managers for its large-scale operation. Products for aviation instruments, altimeters, airspeed indicator and even airplane pitot tubes. Lower NH wage scales helped them win new DOD contracts. Plus, the new plant had a lot of excess capacity for future work. For top manager slots, no director's jobs were available yet. However, one Director was reaching retirement age they were seeking to hire a Manager who would be promoted to the slot in a year or two. A package was like what I was making and with no New Hampshire tax on my pay and a 15-minute commute to work. I was ready to make a change.

I joined Kollsman in September 1977, the winter of the big one! The great storm of 1978! A snowstorm that left wall of snow at the New Hampshire Massachusetts border that was several feet high. So much snow, that no Massachusetts-based employees showed up for work for almost 2 weeks. I, however, was on time every day! I just needed to put chains on the car to get from home and back; otherwise, I never missed a minute on this job. Matrix organization charts were popular in those days. Project Management teams began appearing on functional managerial charts as branches. And all Project Managers and functional Directors at all reported to the head General Manager.

Photo 10.0: US Army M60 Tank...We won the Fire-Control System contract – My fellow MBA and friend Bob Tropea became the Project Manager!

Kollsman Instrument Corporation, Syosset, LI, NY to Merrimac, NH - Finally Director 1967 - Here I was, at 37, going on my 20th year in the electronics arena, on the so-called fast track now, becoming a Director with a former Long Island, NY founded avionics pioneering company...named after the man who invented the airplane altimeter and demonstrated it. On Sept. 24, 1929, by having Jimmy Doolittle flying and landing with a blindfolded cockpit from Mitchell Field in Long Island! In fact, we celebrated the 50th Anniversary of that flight while I was employed there in 1979. I went from a place like Sanders Associates, run by a team of gifted engineering associates...all highly professional players, to join a cutthroat group of survivors of the streets of New York, a union shop no less cutoff by a move to NH! Be careful what you wish for is often said...you might get it!! And I did. I was told by the then VP & GM in charge to take a nasty pill daily if I wanted to get to his or above level. What I didn't reply to him was no thanks, I will keep my soul! I never want to be like you. This guy, with one eye lazy and the other able to scan side to side, an obvious omen to be feared by all, for hundreds of engineers and manufacturing people in NY were fired by him and his cronies. I was like a lamb in a pack of wolves with this group of unsavory characters...just like out of a Guys and Dolls Broadway production. I knew, however, that they needed me and what I could bring to the sandbox. Helping them write a DOD winning proposal worth over $50 million and then my staffing up a team of the best area professionals...with at least seven managers directly reporting to me was a feat not easily accomplished in a short timeframe. I was properly rewarded for my efforts and bringing home 5 figure bonuses at Christmas time certainly helped maintain my wife's lifestyle of fox hunting and socializing with the rich and famous folks of Southern NH at the time. Little did I know that the book "Payton Place" was based on the then fast-lane, Southern NH lifestyle.

After working to win a new contract with a code name of Project WALL, a hugh take away job from a big-league player, we had a celebration party at the local country club with all our proposal team and our spouses. The code name stood for "Winner Take All" and the initial phase one was for almost $100,000,000! Several follow-ons occurred bringing the total contract well into the hundreds of millions.

I recall my first several weeks burning the midnight oil and being assigned as a writer whose specialty was to provide process flow diagrams along with quality control inspection stations strategically placed to optimize manufacturing yields and productivity. Later, my good friend Bob Tropea, became Project Manager and his factory floor was laid out just like our process diagrams described them in our proposal submittal. Looking at our results from a purely technical viewpoint, made our submittal highly probable of success. I doubt if other competitors had anything comparable.

Several months later, in my newly appointed Director's chair along with new large office and high-level furnishings, I was chosen to participate with a new team of chosen managerial players to be sent to a New Jersey military base to negotiate the best and final offer (BAFO).

Most of you folks who are familiar with DOD contracts, recall the final negotiations phase of these, many times, very lucrative military contracts, especially with add-ons and award fees. Before final signing, the military contractor often invites the bidder to a BAFO (Best and Final Offer) where the best price is negotiated.

The Managerial Team Leader was a several Programs Director level bosses known as "Ivan the Terrible!" His nickname came from the Russian Ukraine Province where he was supposedly born and raised. For the record, this guy was one of the nastiest SOBs I ever met in my 50 years in the business! And that is not an understatement! Together with his VP boss, both could act as tour guides in hell for eternity and enjoy every minute of their plight and ordeal!

As I said, this was my first Director level attended BAFO and thank God it was my last! Both sides of the table, spent days knock down, heated arguments, using all forms of profanity, and verbal assaults, personal insults close to physical violence over what turned out to be a 5 to 7-million-dollar incentive award prize. I cannot repeat some of the verbal action used at that negotiation among a mixed gender team of government negotiators who did not refrain Ivan the Terrible from doing his thing to me a low-life act that was embarrassing for me to witness. After the day's meeting, we would dine at the finest places and Ivan continued with self praises of his intimidating performance. It was a nauseating scene to be part of and I could not get over what I had become part of.

Upon returning to our office in NH, Ivan landed in the local hospital for several days for who knows what ailments, for we were never told. Later upon his release, his buddy the VP praised his performance for somehow the government caved and we were rewarded the maximum dollar amount for our efforts. I somehow can only imagine what goes on at higher levels negotiating since the contract involved just a paltry sum of a little over 50 million dollars.

One of my new Manager's, a retired Navy Officer, who everyone liked and had a deep respect for, a straight shooter, discovered some actions violating company policy and wanted to document the infractions. With trepidations, I Ok'd the write-ups after careful consideration. Although we were right, there is a saying, you can be dead right! And we both were! This situation involved another Director who was drinking buddies with the VP who ran the company. As a result, we both were terminated as soon as the headman found out. As a result, later this same Director was promoted to VP, with several other VP announcements, while my guy and I were given our walking papers. Later, the US Government in-plant Representative, who knew me from my Sanders years, shook the trees and many victims resulted from our both disappearing from our workplaces. More serious actions were later taken, some years later, as you will read on.

Two weeks after leaving all my fellow Directors were all elevated to the Vice-President level...and I was on the street after just two years of dedicated service. Jumping ahead to 1985, I was in my corner office, high overlooking Lake Destiny in Maitland, Florida and my Secretary said I had an urgent phone call from NH! I picked it up and this is what I heard to the best of my knowledge. Chuck, I wish you were here now to see what I am witnessing! The NH State Police are escorting your former boss and his Secretary out the front door...under arrest! We just had to let you know about this incident. I hope you are doing well; we all miss your presence here in NH! Here I am in Orlando, with almost 30 of my fellow NH workers there now all here working for what was a Big 10 competitors, now our bread and butter employer. Some say if you live long enough, you will see justice...and in this case, I did! And as a sequel to this story, in 2015, as I am writing this, I read in Florida's Tampa

Sunday paper, with an indifferent attitude, that at 91, living nearby me in Florida, the SOB I worked for and released me in NH passed away last week…so much for that era! Reminded me of the tune that "only the good die young!"

MFE Corporation in Salem, NH is where I landed, just weeks after leaving Kollsman Instrument in Merrimac, NH. This was a business turnaround effort. I kept my Director's title and even increased my basic salary but left for Baltimore to join Martin before all my management colleagues again like before were promoted to VPs and given substantial stock bonuses. This job was a working Director's position, not just a desk job involved in meetings after meetings just to entertain upper levels of management. I even ran the plant on second shift for a few weeks, as a GM, when we needed to meet demanding customers for our new computer hard drives and other products in great demand at the time.

I had all the traditional assurance functions and in addition after a short time took over Customer Service too! This proved to be invaluable for I organized this function to alert management of field problems before they became recalls as most of our served market involved medical electronics products. After taking inventory of returns and analyzing reasons and charting distributions, I gave management their top 10 contributors. We immediately changed factory processes to eliminate most of the heavy hitter returns and profitability improved. In fact, everything we did, the Operations Director and me brought even bigger margins to all our product offerings. I became good friends with the Ops Chief for he taught me many things about manufacturing that helped me in my career down the line. My immediate boss, a GM who was a Harvard MBA, and let us run the show if we made our goals, and we always exceeded them. This job showed me more than ever that factory turnarounds were possible and demonstrated they could be fixed in short order with just few corrections to the process.

At the time, Wang Laboratories and DEC were the new booming companies in the Route 128 high tech arena.

Doctor Wang did one hell of a job riding his new digital calculator mainframe server that was distributed around all offices. This later was the invention that preceded distributed processing with computers that DEC dominated for many years with their VAX Computers. If anyone said that both would be gone by the turn of the Century, we would have laughed at them! But HP bought the company from Compaq that bought DEC and Wang went bust after his son took over and ran down the company in just a few years. Doctor Wang, a Route 128 icon at the time, dying of cancer, even left his hospital bed to try to fix things, a valiant effort, but to no avail, like himself soon it was all over for both. As I learned so often in high technology field, you either grow or die and that is a fact of life in this glamorous business.

11. From Boston Route 128 to Baltimore/DC Beltways

I was 43 when I relocated to the US South again, after spending a few years there on US ARMY active duty in Huntsville, Alabama. This was a big deal for me after working and going to colleges for almost 17 years in the famed Boston 128 high technology area. I got to love my new home and my lifestyle was going well. However, my personal world was in turmoil resulting in a bitter area divorce that I still, on fewer occasions, get nightmares over...I call it post traumatic divorce stress disorder or PTSD squared!

My career and workaholic drive consisted of climbing up the corporate ladder to the Director level of a large company and teaching college on the side. Teaching was a so-called hobby, that I loved doing. When this second job became teaching several classes a week and correcting paperwork on weekends along with my new Director's job also requiring a 10-hour a day commitment, my marriage started to suffer. I recall one night quickly going home for my class notes that I had in several briefcases and arriving in my classroom.... oops...the wrong subject and the wrong night. Well, I did something only seen since in one class at Babson, one taught by Boston renowned Howard Anderson, founder of the Yankee Group in Boston and at the top of his game! He entered the classroom introduced himself by throwing a bunch of his car keys on the desk and did a two-hour standup without any notes to an almost standing ovation audience who sat on the edge of their chairs in awe of what we were witnessing.

Hell, I said, if Howard can do it...I could too! So, I never opened my notes and did a two-hour lecture off the top of my head. Later, after class, one of my best students came up to me and said, Mr. O'Boyle that was the best lecture that I ever have seen...and she meant it!

After that night, many a classes I did without lecture notes...by that time they were memorized, I just looked at my topic outline and went on to give the lessons. My teaching went from two classes, to three or four nights and a Saturday class in all four seasons! I did this for almost 10 years and time seemed to fly by. I recall, after attending college myself on Saturdays for a few years to graduate faster, I promised myself I never again would attend school or teach in summer. But, like a New Year's resolution, this was easy to promise, but so hard to achieve. Especially when all my classrooms, in those hot summer nights, were without the benefits of air conditioning!

On April 1, 1983, I left my Hampton Beach, NH on the Atlantic Ocean beach condo, looked out at the great, beautiful, wide ocean-view scene and said to myself...good-bye New England and never looked back. My Boston years were over, on to a new place called Baltimore, Maryland! And little did I know at the time that I would almost spend 25 years of my work life there! Natives there say one can never leave Bal'More...you're sentenced for life if you come here!!!

Welcome to Bal'More Hon! Was a line one never forgets of the Charm City? After surveying the downtown area several weekends and enjoying the new Inner harbor experience, I was struck by the contrasts of old world and new one so close by. Dilapidated structures such as the old Bromo-Seltzer factory and in nearby Fells Point, large pre-WWII canning factories now empty circled the new Inner Harbor area. One that caught my attention was the old Train Station complex abandoned and now for sale! I asked my soon to be life-long good friend Larry Rosen, CPA to find out about it and let me know what they wanted for it! For I saw renewal projects all over the USA in major city markets and knew of the fortunes that were made there. $275,000 was what he found out! Wow! What a deal, I replied! What are you going to do with it he said? I replied, I don't know but it is a steal!

Later, after not following my intuition and vision, and deeply involved in my new managerial position at Martin, the first part of it went for $200,000,000 now called Camden Yards. The second part another $200,000,000 called Ravens

Stadium! The downtown Chesapeake boys made a fortune! And the rest is history.

Baltimore had a few special things going for it in 1983. The top movie **Diner** was filmed here and debuted at the local York Road classic 1950s theater. And the one time, World Champion Baltimore Colts were still in town, however, attendance was failing. I did get to know and meet football hero Johnny Unitas; a native Pennsylvania guy too, who was an icon in town. Along with Arty Donovan and Cal Ripken, Baltimore was a glorious sports town and still is today. The Baltimore Orioles won both the Pennant and World Series in the year I relocated there. And even today, I still have an addiction the Old Bay seasoning with the famous Chesapeake Bay crabs and delicious crab cakes. I settled down in Baltimore County where I first stayed courtesy of Martin Marietta at the Hunt Valley Marriott, now long gone with the wind too! I stayed at the Marriott so many times during my career that I felt like one of the family. One time in Boston, I was told that there were no rooms left even though I had a reservation. A visit to the Manager's office and after some discussion and showing Bill's personal letters to me, he even gave me Bill Marriott's room for the night. Marriott is a great American corporation and their hotels are among the best in the world.

My first assignment that took about a month to complete was as the writer for the Martin Marietta Baltimore Operations proposal to be included in Martin's new bid on managing the Oak Ridge National Laboratory in Tennessee. This operation housed over 12,000 mostly PHDs in Nuclear Science technology and was a very large facility. After studying Baltimore's experience with portable power supplies made in the 1950s, I became an expert in an area that I always dodged considering the risks associated with this business. I guess I did a good job since later I received an Award for my efforts. Martin won the long-term contract and a guy I met at several MIT technical seminars, when I worked on Route 128 in Boston, became the Project Director on the lucrative contract.

Photo 11.0 OAK RIDGE Proposal Award for my work at MMB…we won the big job!

After a few more weeks on the job at Martin Marietta Baltimore (MMB) I was given another special assignment. I was chosen to accompany the Chinese Trade Delegation on their first visit to the Washington area and was told by management to get to know them and help them in any way I could. It was a Team of 15 to 20 top-ranked Military Officers who were responsible for purchasing all of China's military electronic devices including defense systems. Afterward, I found out that they bought one of everything they could in the USA, it was a very interesting assignment for a 10-day period.

First, they all had on military uniforms, except a few who wore business suits like me. I think that they were the interpreters in suits. I still have their individual business cards Chinese on one side, English on the other...all Directors or Deputy Directors titles! One seemed to be the Leader. He had a young girl who I thought to be his Secretary who stood very close to him at all times. I was very curious who she was since not a word ever left her mouth the entire time of their visit! Later, I found out she had a PHD in Electrical Engineering from MIT, with highest honors, and was up on all the latest US technology. I was amazed dining with them daily and seeing how much food and beer they all consumed in private Asian restaurants all over the Baltimore and DC areas. Every place we dined was closed to the public and only opened for us. It was a very special group of visitors, now customers, that I had the privilege of helping host all of them. My only problem was their continuous smoking and being a former smoker and quitting at 23 when on the package, it said "may cause cancer" appeared the very first time back in 1963. I recall in the Army after many miles of marching, we were given WWII rations of Old Gold or Raleigh brand stale cigarettes and told to "lite up if you had them!"

A little background on MMB before I get into my experiences there. First, my predecessors were now in much higher positions, all insiders and "flameproof" as they called it in Bal'more! One had jumped up two steps in the chain of command to the Director of Procurement. The other, to the job I was eventually to take Director of Quality Assurance. Little did I know before I accepted the new job of the dangers for me that lay ahead.

In my first tour of the large WWII factory frozen in time except for a new PC board stuffing operation for the new US Navy Vertical Launching System Program. And remember that the place was a UAW member union shop too! Hourly paid factory workers often voted to strike and were offset by majority members in Denver and Orlando Operations. Although I had no hourly workers who were union members on my staff, I was still Management from their viewpoint and disliked by the rank and file just because I was a Manager. This situation became clear to me one night when leaving work to go to my car parked in the **"Managers Only"** area. I noticed that my new car was almost touching the ground! Yes, union unrest had hit me too! My new car Michelin radials were flattened by what appeared to be stab wounds perhaps a shop file! I immediately recalled the BW Parkway sign that read "Welcome to Bal'More Hon!" I received my first taste of what was to come at MMB!

One factory operation on my initial tour was an assembly line for welding members of a missile launcher. A completely unmanned process appearing to be fully automated with robot like tooling that placed each extrusion frame welded together and moving through a series of liquid cleaning and rinsing baths. It reminded me of putting together my kid's erector set one Christmas morning years ago and as a kid playing with large structures repeatedly. Foolproof! I was told and no overtime pay for this operation something the old-school players were very proud of! I was not that impressed after touring and auditing some of the best factories in the World. This operation seemed crude and hazardous and I did not feel comfortable with the lack of eyes and ears over the operation. My gut feel was right, this operation running for years before I arrived there became my nemesis for a big career at Martin!

Photo 11.1 MM Corp. Advanced Management Program Training Certificate from Atlanta location.

My first action was to put people on all operations I saw were critical, including a young man hot shot on the launcher program. He was a Chesapeake boy and when the shit hit the fan, no longer was associated with this job assignment, but was moved to another project. With a choice for some down-home stories of hunting and fishing, little else was seen in his performance on the job! Before he left, he recommended another good BS'r who did little to prevent bad product from being produced.

The local Chapter of the ASQC asked me to address a large annual meeting in B'more to talk about the latest craze happening in the USA....Quality! My talk focused on the new opportunities for Quality professionals happening in US industry of taking over the CEO positions in the top Fortune 500 companies. The auto industry was first with Chrysler and continuing in the Defense sector. I guess I hit a new trend and they gave me a very happy reception at that session. My Award plaque is shown below of that important occasion.

11.2 Photo of ASQC Speech Award done in Baltimore at Conference.

I was chosen to put out a few fires in the field at MMB. One started on the shop floor where large staging operations on the fan reverser for jet engines were being built. These parts were placed on tooling that looked like a lazy Susan arrangement where many could be assembled in rapid fashion. I closer look revealed a single casting piece that made up the main part of the assembly. Arriving at work at 7:00AM, factory start time and seldom ended for me until 700PM when I finally left the plant operation! One of my engineers asked me to follow him and I did. Straight to the fan reverser line he took me and said look! Do you see anything wrong? I complied and put on my reading glasses and saw what appeared were what looked like casting marks not anything to get excited about. He took out his pen and said look here! I replied, oh I see! What appeared as a thin (crack?) cut about 4 to 5 inches long near the center of the piece. A flaw in the making of the original part! No need to tell me as a very frequent jet plane flyer what this meant! No stopping a landing jet from this flawed engine reverser! Worse case thoughts crossed my mind when instead of doing its job, directing the jet engine air flow, the part could explode, and a catastrophic failure could occur! I looked around and saw that there were 60 to 75 housings just in this production area that was stopped since the initial finding was discovered by our Resident Government inspector.

My job was to find out what happened and get it fixed …ASAP! It was Friday and my earliest flight to the casting manufacturer was scheduled for Sunday morning. Arriving at the front door, I was greeted by someone whose business card read VP of Marketing! Signing in the Visitor's Log before me on Saturday was our jet engine customer. As a frequent flyer, I felt a little relieved knowing that they were there before me. Since no one was there at work on Sunday, the VP asked me to come back on Monday and we will get down to business.

The next day, I got there bright and early and asked for the Quality Manager who was nowhere in sight! I did see his office the day before close to the production floor. Oh, the VP said, I'm the Quality Manager our guy is out on extended sick leave. Well to my mind, he almost admitted with a straight face that there were troubles right here in River City and that he was part of the problem, if not the problem! I asked to be taken to the Inspection Department area and the Final Shop Inspection Station. Oh, he replied, they are one in the same! Some castings were on the production line but exhibited none of the dangerous looking slits that showed up on the MMB Assembly line. In fact, upon further checking on my part, none of the operations here had any of the suspected parts.

I asked for an Inventory Control dump (what was counted at last inventory time, usually done once a year in most factories, once a month in others). Wow! 1500 counted here somewhere in your stock inventory area. I asked for a sample of that inventory be displayed forthwith. He said ok we will pull it from the end boxes. No, I replied! I'll pick them and did a random sample. That will take some time and need a forklift, he cautioned. I have the time, no rush, pick a few from these boxes and I'll stand by and let's look at them! He looked a little startled and said, ok, you're the boss!

I won't bore you with all the details, his sample had "zero defects", my random sample had 100% defects! Now to the root cause of the problem! It seems that recently MMB approved a stamping tool rework using what is called "shims" that was quite unreliable and was the cause of the defective products we had received and were in inventory stock here. I requested the part drawings and noticed that no one signed off the so-called tool rework fix! Later, back at MMB, I found out that one of our buyers gave his blessing to the change that saved thousands of dollars and many work weeks to make a new molding die tool. This is how accidents happen and sometimes undetected root causes are never revealed. It was a few months until production resumed with these fan reversers and as far as I knew no one was held accountable for what could have been disastrous results had it not been discovered. All the defective parts were scrapped, and new replacements were made at no charge to MMB.

Another war story from my time at MMB was a troubleshooting trip I was to make to an electronic connector factory in mid-west USA. We had some of their products that appeared to exhibit green residue like mold discoloration on our MMB electronics production assembly line. This part was mil-spec level (military specification produced) and was used universally throughout the entire avionics/aerospace industry marketplace! I arrived after several small airplane trips at a remote boondock town and was greeted by the factory General Manager at the front door. I used these parts in many applications over the years and interested in where and how they were produced.

Upon arrival inside the plant, the GM wanted me to be introduced to our in-plant inspector who was here for a few years. I looked around and could not find him as he was dressed as one of their assembly line personnel (Red Flag

No. 3). It seemed that he was having an affair with one of their assembly line girls, but I was not to know this, as I found out later by the GM. I looked around and to my surprise the entire production line was completely out of control! All the process checking control logs had not been signed in months and were out of date. Shop logs were out of date and inspections not performed on all plating and rinsing operations were filthy! The entire factory was in the same condition. Just what was going on here with several other resident inspectors even government inspectors too!

Well our guy went native, as did all the others, this place needed to be cleaned up and fast! When I was finished with my assessment of the place, I gave an overview of my findings to the GM who was very upset with me! "Do you know who I am?" he yelled! "My brother is a top executive with your Orlando Operation, I'll have your job!" He went on and on! I phoned in my findings to MMB and was told to get on a plane and return ASAP! Which I did!

Upon my arrival in Bal'More, my boss told me several firings had taken place out there. The resident and his regional boss were now gone, and we expect several inside the factory would be fired too! I was lucky to escape the place alive he told me with a serious tone! Later, I found out the company I visited cleaned house and all our parts were replaced with new ones at no charge! However, I was never thanked for a job well done, for I wondered if the good old boy network from Orlando was getting even with me for just doing my job!

I myself had a good friend from my NH days who was an MMO Director level executive. On one of my trips, I called on him. He knew the inside story on all issues and invited me to come see him. I always respected him and knew he was a square shooter. His large executive office was in the new Corporate Tower several stories up and in the corner of the building…remember RHIP! Wow! I thought rank has its privilege in force here from such accommodations! His Secretaries office was bigger and plusher than the GM's in MMB! And with several leather couches made his office like

125

the POTUS at the White House in DC! He looked lean and CEO-like as he welcomed me to MMO. Well, how the hell are you doing Chuck? I told you to come on board with the rest of us from NH several years ago. Don't remind me, I thought back to that night in NH years ago when I received a phone call from a headhunter from Connecticut with an offer to join MMO as a Manager. Hindsight is 20/20 I said!

The phone call was made to 30 other people who worked on all the proposals that we soon won in NH taking away all the millions of dollars of contracts from Martin! They got even and one up too! They hired all the proposal team even security personnel and even the Director of Personnel too! That's what DC Beltway power can do in the good old USA! Money also helps as all were given double digit raises and free Florida sunshine instead of NH snow every winter!

My old Director and colleague friend eventually became a CEO after he too was sacrificed and let go. It seemed that one of the top VPs hired earlier from a Bal'More/DC competitor got someone very upset after not meeting his goals and was immediately let go! Some said he liked to be on Sunday morning TV talk shows and was spending too much time off the job he was hired for. Well, all Managers who he hired over the years, including my friend, were cast aside! I heard the number was 20 high-level executives...all managers!!

One job opportunity I was told under his direction was as head of the Missile Testing Program. Another was a new venture called MMMS based in a new facility in nearby Maitland, Fl. I eventually chose this one after my MMB tenure. I chose it because it was a new venture and would be in on the ground floor as Systems Engineering Manager. He told me that he would make a phone call for an introduction. After he gave me a name and address outside then city in a bedroom community called Maitland. I took the job after a great interview with Directors of the new business.

As I looked out of my new office high above Lake Destiny, it was a scene out of a travel magazine. What a difference between this and my windowless office in Bal'More! This was unbelievable, all the best furniture right down to my own credenza minus the duck decoy found in the other high level MMB offices! I went from heaven to hell in moving to Bal'more and back to heaven in under two-years' time!

I noticed a lot of meetings were happening and a very loose organization. The President was a Boston MBA like me who was chosen for his connections and social circles membership. I even was given his new private condo to live in until I got settled in my new job. Later, I found out many condos were owned by other MMO Directors and many rented them out as landlords. Oh yes, my home for 6 months was the concierge level at the Sheraton Hotel across the street from my office. That included all my meals and the stay was billed directly to the company.

MMMS prior to my coming on board, was trying to sell a product concept developed in-house called Remote Automatic Calibration (RAC). This was during the DEC PDP Era and involved replacing the need for original equipment manufacturer maintenance manuals and in place was a nice digital computer screen replacing the need for all paperwork to do maintenance and calibration of test instrumentation. Again, a nice concept, however, since Martin refused to give up their source code most potential customers shied away from buying it! Attending many electronic shows at the time, I developed a new application called Computer-aided Laboratory Automation emulating new concepts seen at these shows and visualizing a broader, world-wide market for my idea. This eventually was to become a billion-dollar business dominated by then IBM!

Upon showing my new graphic marketing brochure to one of the inventors, he replied, "all right this is exactly what we want to do!" However, it was a little late for my idea. I found out the management team already spent over $8,000,000 touring the planet, some with their wives, before I even came on board. This place was a ticking time-bomb ready to be eliminated soon if things did not take off! Later, ahead of my story, my old GM from MMB now VP of MMO and head of this Board of Directors of this new venture, shut down the spinoff and cut all the losses for Martin Marietta Corporation.

It took almost over one year and a half before that occurred, so let me tell you some of my experiences at MMMS. The President of the new MMMS venture invited me to his home near a golf course complex to pick up the key to his condo that I insisted I pay $500 a month for. It was a beautifully furnished place and saved me about $1,000 a month to live elsewhere at that level.

On the way to his home, I stopped by an empty McMansion on the adjacent lake nearby. No price was on the sign just the name 'Husky Realty"! This name and signs were all over the Orlando area for the owner, Vernon Husky, was one of the biggest realtors in Central Florida at the time. My new bosses' home was the typical golf course mansion fitting for one of his business level. I asked him about the home down the block and he replied, "oh, Chuck, that's way out of your pay bracket, they want $275,000 for it!" In hindsight, again, I thought as Hillary Clinton once said, "Coulda, Woulda, Shoulda" bought it!

Jumping ahead, to Christmas time season, I was invited to an Orlando area VIP Party, accompanying my new girlfriend whose Attorney boss's family was an original developer of the new Orlando, settling here in the early 1950s from Philadelphia. Ok, guess where the party was held? Yes, your getting good at this from reading my book! At that same house that was too expensive for me to buy!

Upon entering it, with my stunning date whose Christmas dress just glittered in the well lite home entrance foyer. The owner appeared none other than Charles Givens (later to also be a Club Fed member convicted of mail fraud)! He welcomed me and then accidently spilled his Champagne all over my dates beautiful dress. Oh my, he said, so sorry, anything in my house is yours tonight just take your pick! So, given the opportunity, I seized the day and grabbed his girlfriend, a beautiful tall blond, just to see his reaction! Oh

NO, not that, anything else, but not that and we all had a great laugh! He grabbed my arm and said as he escorted us throughout the house, "we Chucks are all the same aren't we"!

Years later, I read in the Orlando Sentinel that he sued this girl for taking a house he gave her and giving it to her boyfriend! In that home where we were given the 25-cent tour, was everything a man could desire and the view from the top master bedroom floor was marvelous…like out of an ancient Roman garden, lighted pool, lakeside harbor even statues and all! Oh, he also told me he bought it for a bargain…around $200,000 on foreclosure! I almost cried upon hearing that story!

That special night in Florida I met the Who's, Who of Central Florida even the real estate baron Vernon Husky! When I first shook his hand on the greeting line, I told him that he looked like a Real Estate Icon in New Hampshire, who was very successful too and lived down the street from my mini-farm outside Nashua. He laughed and said I sure may need to look him up, we may be related! The party was top shelf with everything on could desire including the live entertainment following diner, nothing less than a New Orleans Jazz Band for dancing and post diner drinks and dessert!

Later that next year, I was invited to my girlfriend's lawyer boss's 37th birthday party. We also went to that special affair and I ended up in the Kitchen with his Italian Parents and Grand Parents, originally from Philadelphia, PA! They were at work cooking up Italian homemade treats that I grew up with! I was welcomed like one of the family. In addition, the party was catered with a soup to nuts array of food and drinks. The party host took me out to his garage to show me his birthday gifts: a red convertible Italian Sports car in one stall and a pyramid of cases of booze going up to the ceiling! Wow, I said, Yes, he replied, my Father's friends all sent them over. Such was inside Orlando in Circa 1980s, the rich got richer and so did all these folks who mingled with them. And this was before Universal and MGM decided to expand there.... which brought in millions and millions more of California dollars to the booming city empire.

Years later, when I returned to Baltimore in 1988 after a two-year absence, I moved to Columbia, for only a short while. And, since I had trouble even finding a gas station there with their so-called Green Environment, I moved back again to Baltimore County. After 10 years and a federal government lawsuit against Baltimore County trying to move over 3,000 inner-city families there. And after the US Department of Health, Education and Welfare (HEW) Director was forced to resign, the county accepted a trade-off and let 300 families move in. The first wave arrived at my remote suburb community and on that first weekend almost burned down the entire complex! This happened after someone's candle lighting ceremony with illegal drugs ignited an adjacent building. Even my place needed to be evacuated and was smoke cleared after the fire was put out by an aggressive local fire department. My next move was to get closer and to a safer neighborhood with less commuting time to and from NASA Goddard. One of my good Baltimore friends recommended a place next to his, a nice Jewish community, just north of the downtown area. I stayed there until my first retirement in 2000. Upon return to Goddard, I decided to move much closer to the base and settled in the townhouse in the town of Bowie, MD, a DC, worker bee, beltway bedroom community.

The only time I felt threatened there was on 911 and when two Beltway shooters terrified the entire DC area with random shootings, I think 2006. Some innocent children at the local schoolyard in Bowie were shot and even a retiree from Goddard was killed coming out of a local Home Depot. The puzzle was solved when at a MD truck stop the suspicious vehicle with a gun hole out the trunk was called in to the local police. I think the driver received $1,000,000 reward for his efforts. I found it amazing what two shooters could do to a large, metropolitan city. The DC area was one place that I did not miss after both of these tragic events.

Photo 11.3. MMB for years, and in 1983, built the launch canister platforms for the Patriot Missile…soon to be my nemesis!

It was now 1983 and the last time I was in Baltimore, except for my interview day with Martin, was 23 years ago, in 1960, and it was at the city bus station on my way to Huntsville! In those days, it was a 36-hour bus trip that I only did once and thank God for that! Baltimore frightened me then at just 18, and the bus station was not on the best scenic tour route. I did not like it and was glad to see it in the bus rear view mirror disappearing. Little did I realize that I would, later, spend almost 25 years of my life there!

Martin Marietta Baltimore factory operation included missile and space products such as the Titan Missile, Vertical Launching System, a NAVY ship missile launching system, Patriot Missile Launchers, aviation sub-systems and associated jet engine parts. The factory to me, familiar by now with all the latest state-of-the-art manufacturing, seem to be frozen in the 1950s or perhaps the 1940s.

131

My department was responsible for collecting all factory inspection reports including a backlog of thousands of deficiency reports. The work required months of laborious inputting of data alone and analysis of the entire factory's product quality levels. Martin used in its factory operation a manual paper system centered on what they called a MARS tag. It went back to the early days of aircraft manufacturing as a means of keeping inspection records. What was lacking wasn't only the lack of an up-to-date data collection system, but not even a computerized one.

After selling the management on using new HP computers, we contracted HP to come in and do a turnkey computerized system. In addition, I hired a team of computer people and formed a new team to fix the long-broken process. This to me was my most important job, for if as managers, we did not know where our problems were and what the magnitude was, how would we ever be able fix them!

Photo 11.4. First B1 Bomber – MMB built the B1B tail section in its 50-year-old facility

In addition, I organized the existing resources into functional teams while trying to match skills, backgrounds and experience. What I found out, at once, that the Peter Principle permeated the entire operation. You had people, except fellow managers, who had no formal training beyond high school or military schools! It was not like Boston, where everyone had degrees from the best academic institutions. Here I was now in, the supposedly, major leagues and surrounded by good old boy appointed players, not experts, but in most cases novices in their chosen fields!

This situation soon explained to me why Martin, once, a leader in its field, has fallen to just an also-ran competitor. In comparison with the other operations, Denver and Orlando was where, it seemed, that all the talent was located. In fact, at union negotiation time, Baltimore always voted to strike offset by Denver and Orlando employees by voting not to strike. After all, the top CEO had a PhD in Electrical Engineering from Princeton University with high honors, not too shabby qualifications. From my perspective, the only thing he offered, besides his DOD & DC connections, was to write his classic beltway bandits' book sheltered from all the chaos of factory operations. I understand, after receiving multi-millions from the aerospace industry consolidation (downsizing), and a free get out of jail card for his efforts, he became a teacher at his alma mater!

In about a few weeks, I had my hands full trying to hit all the homeruns myself with little or no support; it was the blind leading the blind management haven. Some people thought that just going to work was doing their job, and once there, socializing with the good old boy network on how many ducks they hit last time hunting on the Chesapeake waters. MMB was a scary place to be in 1983, unless you were a high-level general manager groomed to soon be the new CEO.

Photo 11.5. Martin seaplane built in the Middle River Plant in the 1930s

Pan Am Airlines used the Martin Clipper ships in the first USA passenger flights in the early '30s! Don't get me wrong, after all Glenn L Martin taught William E. Boeing how to fly! And look at both companies now to see them from my point of view as a manager. When I was there, the World War II/Korean War vets were still hanging on, although all heavy smokers as that entire generation were, but their days were numbered. The original program manager for the first missiles, the Titan was still on board. A quite likable good old boy he has spent most of his life working in that facility. And the B-1 B Program Manager was also from that greatest generation era and they were the leftovers from the Glenn L Martin glory days. I recalled that Martin had a close call in the 1950s, when the merger with Marietta saved the company from going under. And the guy named Bunker miraculously made the leap from airplanes to guided missiles. One story that I must tell you was right after World War II the top Navy Pentagon brass was invited to Martin's airport, on the same location as their factory, to witness a new product called a jet. After the successful demonstration, the top brass concluded with this famous Martin line, "If you can slow it down, we may be able to possibly use it"! So much for that advanced Annapolis training vision at the time!

Also note that during World War II, the US Naval Attaché at Martin would eventually become the President of the United States of America. None other than Richard M. Nixon, US Naval Officer Attaché! Also, after leaving the US Vice-Presidency, Spiro T Agnew also rented office space adjacent to the main complex. It was an office that I frequently liked to visit after his departure; especially seeing the soundproof conference room adjacent to his office. It certainly was hallowed ground as the President of the company mentioned in the speech, in 1984, at an all hands meeting outside on a nice Baltimore weather day. He also dedicated a new engineering office building to be erected on the premises. Today, it remains as the last of the glory days for Martin's iconic facility.

My tour of duty at MMB was only about two years before the shit hit the fan as they say there! As I mentioned before, the big leagues are ruthless in finding a blame for any loss or major incident. As I was soon to find out after finally getting things staffed and properly organized. As it happened long before my tenure, one product line produced then suffered serious field problems with a rocket canister made in MMB. The root cause was found out to be a so-called automated process that was unmonitored and seemingly out of control for several lots of production. This resulted in crews of technicians being assigned to fix products that were deployed all over the world. And costing the company thousands of dollars to clean up the problem. Not only to mention the PR damage that suffered by producing such a problem. Since I was the new kid on the block, and the Quality Assurance Department was now under my watch, not my boss, the Director, who occupied the role prior to me or another former boss, now promoted to Procurement Director and who had my position for years before him!

I was placed "on staff" as the term for removing a manager was given in those days and another new hire player was given my position. And since I was truly innocent of any involvement, I was given the opportunity to transfer elsewhere to Orlando (MMO) or Denver (MMD), where they could make good use of my talents. After several in house interviews, I chose Orlando at a new spinoff company that seemed a bigger challenge to me. Especially seeing that my former MMB GM was now running all MMO Operations as a new VP and was soon to be the new CEO!

I found if amazing how these guys could be promoted despite disasters occurring all around them. They were the "chosen' ones and almost Teflon-proof from such incidents. I took on the new assignment as Systems Engineering Manager, a top-level job at the new start-up. Little did I know then that the top management team, of this new spinoff, were on the ropes after spending millions on international travel and business entertainment that yielded little or no sales revenue. Later, I was told that the T & E figure was over $8 million dollars.

I wore several hats there. First, I put my MBA marketing hat on and developed a product line using my HP model that I was very familiar with. Next, I created new product applications literature and a brochure that according to another seasoned manager "hit the nail right on the head!" I followed by writing up a Marketing Plan called "Bowash" to go to our served target markets of big Boston to Washington companies who would buy our offerings once convinced of their payback merits! And then actually travelling with our Marketing Manager to all of them to show and tell our story. The only problem that I was soon to learn was that they, now we, were selling an unproven concept with little substance; somewhat like the Las Vegas smoke and mirrors of magic acts where are all so familiar with. The new idea needed substance; not just a library of software applications focused on commercial test instrumentation built by leading companies such as HP and Tektronix along with a Digital Equipment Corporation VAX mini-computer. This system called Remote Automatic Calibration was to allow quick checking of electronic equipment's calibration and overall performance without the use of manuals, all using what looked like a desk top computer of its day.

Project CILO (Computer Integrated Laboratory Operations) was a new proposal out of NASA's Michoud, LA Facility where the Space Shuttle External Tank was made. My first visit in July gave me a good reason never to visit New Orleans in Summer! After opening the rental car door all windows fogged up with steam. The temperature was in the 90s and the humidity was at that level too! Luckily for me, I wore my Palm Beach Summer suit as it fit with the day's weather.

Michoud was a government owned facility from the days of NASA Saturn space missions. Today, however, it produced the famous External Tank, a rocket type booster that catapulted the space shuttle into Earth orbit. Quite a feat of USA engineering as each one made travelled by barge down the River ending up at Cape Canaveral, Florida. I got the 25 cents tour of the facility and we were on a sales tour to bid a new NASA Computer System buy called CILO. I ended up making several trips to Michoud before they ended up giving the job to IBM to avoid a conflict of interest charge since we were all part of Martin Marietta…as they held the facility contract at the time. I really enjoyed working with the Michaud Team some even had PHDs in science and were experts in their fields required by such a feat of technology. On one weekend, we stayed over to see the New Orleans Jazz Festival, one special scene that I was so lucky to see.

Back to MMMS, in addition to my Systems Engineering responsibilities, I was given the job to manage a new US Army Project to build a complete system for them to evaluate. This system was to help increase the uptime on overseas base deployments with high maintenance intervals and to optimize systems readiness. Assigned to me, as Quality Manager, was an old Pearl Harbor survivor still working as a reward for his military service. He was a nice man that I respected for his involvement in WWII, but knew little about the technology, let alone the art and science of the Quality Assurance business.

Well to make a long story short, the Army sent the system to Tobyhanna Army Depot close to my hometown in PA and they essentially "tore it to pieces"! Perhaps my old high school buddies who worked there even had a hand in the systems evaluation. However, a failing report soon followed that ended this, what would have been a lucrative Laboratory Automation billion-dollar world market for the new technology at the time. Top management at MMO, who made up the Board of Directors of the new venture, soon stopped the funding of the new spinoff. And I was put on staff again, a support position without any responsibility. My thoughts then were to return to the Northeast, as I missed my family in PA and I started looking to leave Martin where I came to dislike, after nearly 5 years of hard work and unrecognized service.

A few years later, an old employer called Lockheed, eliminated a competitor, by purchasing Martin and moving 350 executives and their support team to Bethesda, MD, where to this day, Lockheed is doing a bang-up job running a highly profitable world class organization.

Martin and Lockheed were a different "breed of cats" as I called them. Lockheed was a west coaster, more businesslike, more professional and as both stocks showed, better managed too! Although Martin's stock seems to plateau for years, prior to the Lockheed merger. Had Lockheed not merged, I think Martin Orlando and Denver operations would have survived. I doubt, however, that the Baltimore Division could keep up with the leaps forward in both technology and profitability.

Wanted: Director of Command, Control & Communication in Virginia Beach, VA! The ad at the Orlando Marriot Hotel read where I had just had a TGIF lunch! Yes, I said to myself I can do that job! I went to the room where they were interviewing and given the ok to go to Virginia for final screening. I passed with flying colors and went shopping for some Director level dress suits with vests and suspenders to fit my new leadership role! Dressing for success was always a pre-requisite for successfully assimilating into a new organizational culture. Just after a few weeks on staff at MMO, I was in the arena again slaying the many dragons associated with a high-level role. Upon joining the new organization, I found out that I was replacing as two-star, retired US Army General, who was not familiar with the hands-on details of systems engineering projects. I had two programs directly reporting to me. Both Program Managers were ex-USA Naval officers. In fact, the company was filled with retired Navy folks since the large US Navy base at Norfolk was nearby.

One of the biggest secrets that I discovered in the Baltimore/Washington Beltways was that resumes, and credentials count a lot. That's where people like me were sought out as "hired guns"! Having one's name in a proposal submittal package or on a company organization chart can result in millions or perhaps billions of dollars of revenue for a Beltway Bandit. For example, my resume with the best of Boston education, graduate school teaching credentials, Director titles were invaluable in receiving and retaining US Federal Government contracts! In a Pre-Award Survey by a DOD Agency, the military representative said, and I quote "I am impressed by the level and top-notch personnel assigned to this contract"! And he went on to approve my contractor's proposal submittal.

Minority owned businesses flourished during my 25 years inside the Beltway corridor and Bal'More Beltway too! They often say that the government does not start-up businesses but, in my time, I saw hundreds of new enterprises spring up in just one region of the country. Millionaires were made perhaps not billionaires as was Bill Gates and Steve Jobs but very rich players in a wide-open Government marketplace. My NASA colleague Joe always commented that he even as a Navy Veteran and expert in his engineering field would also become quite wealthy if only his little self-owned business received just 10% of the attention of these new players. As I mentioned, I too had become a hired gun for two of these new 8A organizations. Continuing from earlier text, this was my first and it was not a pleasant experience.

The first program involved the modernization of the entire FAA Airport Control Tower System. Pilot sites were in nearby Atlantic City, New Jersey. The installation of new large screen computer airline tracking systems that looked like the large high-definition TV we use today with the finest screen resolution was a big improvement over antiquated terminals of the 1960s. The company progress for government monitoring in this contract was awarded to none other than my former employer Martin Marietta. Obvious now why they were in Orlando to find someone to be in charge and now that guy was me!!

The second program was a US Navy contract to upgrade the fleet with new, ruggedized personal computers for deployment all over the world. I was responsible for planning, organizing, staffing, controlling and directing both large scale programs and felt like a quarterback calling the plays trained and experienced in the science and art of management. The only tradeoff was that my personal life suffered in that my workday provided little time for any R&R activities. After just a few months, six to be exact, I opened my morning newspaper and the front-page headlines read **"Local CEO Staff Indicted"**! And guess what? Yes, it was my new employer…who, at one time, was SBA Man of the Year! And on our first meeting, I recall seeing him in a photo of the President Ronald Reagan shaking hands as you entered his office suite! Several members of his staff were indicted! So much for that next rung up the ladder in my career! The CEO's cousin, a Sr. Ex. VP, called me into his office that morning and said that he regretted to tell me that the US Navy had pulled the plug on funding and they were retrenching the organization! And I was not part of their downsizing plan. I left Virginia in a few days after storing all my belongings and never looked back on that assignment other than refreshing my memory for this book on that somewhat trying time in my career.

Growing up, I always felt that I had a specific purpose to lead a large organization and use all my God-given and earthly talents to achieve a difficult and somewhat impossible task. This drive soon called me to a place that I could accomplish this repeatedly, but it was not in Virginia.

My last shot before finding my NASA niche was with a small DC beltway firm founded by a Viet Nam Era Army Commander that received a US Army 8A Set-Aside Contract to build a mobile, battle-field deployable medical records communications system. They needed a Systems Engineering Manager to get past PDR & CDR hurdles (Design Reviews). This was prior to Desert Storm and what they were trying to build, knowing the large number of potential casualties projected to result from the conflict, was a battle-field deployed computer system that could be used up front to keep track of victim's medical records closest to the point of occurrence. Like so many of these 8A start-ups, a high learning curve existed and there was little time to train inexperienced personnel in the complicated tasks and little or no funding anyway, if they had the time. I was told later that even IBM could not get the systems built in the allotted timeframe since the war was eminent and these systems would be sorely needed!

The Project had several consultants who were milking the job as I was soon to find out and delaying progress if they could. I soon found out that the term consultant had a double meaning as did a lot of beltway culture did. Their job here was to expand the task and stay on the payroll if possible! They were paid both here and from the companies they represented whose hardware and software was part of the requirements! Only Beltway Bandits could pull off such a scam in my estimation. I did my best to use current resources later found out to be contractors with good credentials, however, like consultants averse to doing any work of value. Many had BSEE and MSEE's from Southern Colleges and Universities not like Boston folks, where I was always impressed by the level of talent and expertise. But that was in the 60s and 70s this now was the 1980s and the big demand degree was in Computer Science although EEs for doing the work I needed were of great value. If they could work and contribute, even more they gained my respect. After working there several months working harder and longer than I ever did, including Saturdays and Sundays I felt the weight of the world was on my shoulders…and there were no more rabbits in my big hat!

I spent over 6 months trying to get the place organized, staffed and in control. However, what was needed were many highly experienced communications skilled technicians working 24/7 to build the system that would work and meet contract requirements. One day, I was called in to the Program Managers office and told "meet your new boss!" He was a recently retired US Army Signal School Colonel from Fort Monmouth, who was taking over the job. He was to be calling the shots and I would be reporting to him! Well, as they say, the writing was on the wall, two weeks later I was given my final paycheck and it was back to PA for me! Here I was at 48 years old without a job, but a AS, BS & MBA and with 30-years hands-on experience in high technology!

My Dad was a great motivator, an expert in using both positive and negative techniques as applicable to the situation. I think his negative technique worked on me better than he thought. For example, after my speedbumps getting back on my feet up north again, he planted a seed that bore results in short order! Telling someone it can't be done or that you can't do it is sometimes more powerful driver than just saying otherwise. Reverse psychology some say that it is. And Dad knew what drove me as a caring Father should. "You are over the hill now he would say. No more big-league jobs for you, you are too old! They can hire two young guys for your salary!" Always planting these seeds into my brain…for he knew it would work! He seemed never satisfied with any of my career attainments in life and had a deep-seated selfishness of wanting me by his side as I did all through my childhood years. "Why aren't you head of NASA", was one of his favorite digs. But he was my dear friend and our best times later in our lives was watching the great Joe Paterno led Penn State games on Fall Saturdays having a bomber bottle of Colt 45 and local delicious pizza bought at senior citizen price of $6.00 and beer cost of $1.00 a bottle! Yes. For a total of $8.00 (two beers included) we enjoyed a priceless afternoon of college football. And if Penn State won no day could be more perfect on this planet!

Dad played in the same high school as I later attended, and he was a member of what was called the "Scoreless Wonders" who lost every game by very low scores of 2-0 and 6-0 all single digit losses. My Dad's younger brother Uncle Billy, scored the first touchdown the next year and played with the NFL great Charlie Trippi, an icon born in my hometown of PA. Charlie as of this writing is still alive in his 90s and living in Georgia, his college alma mater state.

Within weeks of being let go in Virginia, however, I was back in the game again in a job now owned by the minority owned business entrepreneurs. The good old boy networks around DC are many. First, you have the ex-govies, then ex-military, then ex-contractor management who spin off and do their own thing.

Photo 11.6. Huntsville, AL in 2009 at the new Space Museum and an interested tourist (me) who used to live and

work here! Note the large, Saturn rocket standing in background!

After retiring I took a drive through the South to see all the changes. Returning to Huntsville on my 50th Anniversary of being first stationed there, I was amazed by what I was about to see. It was not what I saw in the 1980's on my many business-related trips to Huntsville. I recall on one occasion; I revisited the old Space Museum to again see Dr. von Braun's personal office on display and it was gone! Puzzled and very disappointed, I went over to an elderly guard and asked what happened to Dr. von Braun's office…desk, chair and all? Gone now, he replied! Why? I said. He went on to say that the Space Program was now going to emphasize American history only, not the German influence that was formerly presented. I shook my head in dismay…another experiment in social engineering…Von Braun was a citizen and most of the 165 German Scientist were too…why were they now being dishonored, why? I wanted to look again at Dr. von Braun's books he had written that I on earlier visits held in my hands. I wanted to see his first drawings of his space plane…. that had wings…like later shuttles designed by Rockwell International in California. The drawings were so futuristic; like the ones I saw in my early childhood year's comic books. And I wanted to review the many Doctor degrees earned and given in his honor, literally hundreds, and now 20th Century history was gone with the wind as they say!

Photo 11.7: Dr. Werner Von Braun's Office preserved today in the new Space Museum in Huntsville, Alabama.

Today, however, most of his things are now restored to their proper place in the new Space Museum, now a fitting place for such treasures as well as other historical collectibles. One I vividly recall was what I call the umbilical cord of the Saturn Missile...a large bundle of cabling, like human veins that powered the spacecraft...so artfully made and assembled by IBM Corporation. The highest quality I ever saw done by American industry craftsmen...second to none!

Welcome to Goddard, the sign said as I entered my first workday. Robert Goddard was a scientist, and some say the Father of the American Rocketry. For in 1918, WWI, he was working with the US Army Signal Corp trying to find methods to reach extreme altitudes using rockets. While we were still at war, new technology was needed to keep up with our enemy's advances. US Army Ordnance goes back to 1812 and continued to in my time with Explorer I. An excellent book covering Rocket history is Mike Gruntman's book entitled ***"Early History of Spacecraft and Rocketry"*** published by the AIAA in 2004 out of Reston, VA

One of NASA's most reliable launch rocket then and now is the Delta, an on-going Program based in LA, CA at McDonnell Douglas (MDAC). My first of many visits to MDAC California facility where the Delta missile (since the 1960s) is made was to serve two purposes. First, to meet my on-site Team Leader, who was paid more than me as a sub-contractor! And secondly, to walk through the facility to make sure that we covered all that NASA expected and that I too expected. Well both purposes were successfully fulfilled. After meeting Leroy (Lee) Geisel, a Holy Cross, MA engineering graduate who for his entire career worked here (35 years) I understood why he was paid so well!

Having many years' experience, myself with Aerospace operations, I told him, take me around and I'll tell you if I need some additional explaining on the tour. I told him let's start at the receiving dock and end at the shipping dock. He said follow me. Well, in 45 minutes later, we completed our factory tour. Lee did an excellent job of showing me the facility. I was immediately impressed with his intimate knowledge. When we walked to his big office…next to the Plant Managers, I realized that here was a very special engineer…respected by MDAC both as an ex-employee and as a hands-contractor. Here was one place I could trust, and never did I need to worry about this critical job Lee was doing. Delta then and today had a highly successful record almost setting the bar high for others to achieve.

Along with Lee, I needed to find more experienced players to backfill in chance Lee retired. My chance came when in San Diego General Dynamics (GD) announced a complete shutdown of their facility there. On to San Diego I went. GD Personnel set up an all-day session for me to interview personnel who were being let go. After what seemed hundreds of interviews, I luckily found three people later to be my best performers, to immediately hire. One is now with his own company and still covers all the Delta launches. Another went on to be a Director of the DOD's Missile Defense team and is now in the Senior Executive Service in DC. I met him years later accompanying a several stars Air

Force General and he said that it was all my fault that he got there. I laughed and replied thank you...you were always one of my favorites!

The USA's high technology business is not as big as one may think! Steve Jobs couldn't get a full-time job at HP because he did not have a Stanford EE degree. They told him that...one of their big blunders...now his Apple is bigger than them all. Similar stories are abounding in this fast-moving grow or die industry.

Bill Gates appeared in the Boston Globe in the early 1970s, with a then looked like a motley crew photo announcing the birth of a new business called Microsoft. From the photo, that, looked like a wanted poster...one asked, do I invest my money in this new company? Not me, I said, regrettably today.

In Orlando, years later with several of my engineering friends, we pondered about the new internet...it was 1987 or 1988. How can we make money with it...any ideas? The answer was unanimous...no way, it's free! How can anyone make money on something that is free! So much for that gang of non-entrepreneurs! Years later, we all thought about all the new young billionaires in Silicon Valley...my best friend, Frank, an EE, now on a HP pension, said that we were having too much fun anyway...that was our tradeoff!

Let's see what we missed from that brainstorming session of the 1980s? Facebook, soon to be the World's first trillion-dollar business, just announced a $1.65 Billion profitable Quarter...think about that over $500 million a month in profits! Google who would have thought that funny name would buy out Motorola, a giant tech house around since the 1940s! And who ever thought of the I Store that revolutionized the music and entertainment business along with Netflix? And your ISP that you take for granted and pay dearly monthly for that fast speed! And who ever heard of paying hundreds for just using a mobile phone? I still remember my parents party line number in the 1940s...2641-R; when we could listen to all the

147

neighborhood gossip before the social media was even dreamed about! Our bill was under 5.00 a month and whoa be tied of you if you used a long-distance call! Our trick in those days to phone in safe arrival on trips was to call collect for Uncle Harry…person to person…and they would answer… "he's not home, but he will be back later!" We all used that method for years and years! I will not bore you with more internet success such as Amazon, Yahoo, YouTube, Twitter, Netflix and now the Cloud!

However, it's a good time to look at where we are today with our PCs vs just 20 years ago. Back then is when the Fed Gov forced us to use Microsoft technology in the early 1990s. If you wanted to bid on a new proposal for government work this was the requirement. At the time, I was using Word Perfect and Lotus 1,2,3! In place of what we thought was superior to Bill Gates' software! Meanwhile at HP and Apple, high technology gurus were hard at work designing new personal computer products. At Apple engineers were designing in-house with what was called closed-end software design. As a result, even today, Apple computers, using this technique of designing in barriers, have a high degree of protection from internet threats.

Meanwhile at HP, PC hardware engineers were using Microsoft's Operating System that even today, is highly susceptible to internet threats such as virus, malware, Trojan horses and a myriad of open-ended hostile software bugs! So much of these exist that now a billion-dollar business has been created and is still growing considering the high degree of cyberwarfare taking place. Even when you purchase a new computer, they give you a free trial of so-called virus protective software and add-on intervention programs and some not even made in the USA! From my point of view, any need for extra protection is a product design void and what we used to call a plain old "Computer Bug!" back in the day!

Today, the computer threats are higher than ever, and now foreign nations, in some cases, have superior skills to what we possess in the USA! Our USA Colleges and Universities are, in many cases, teaching obsolete computer course offerings! Some go back to the 1980s, when I left the college lecturing circuit and are 20 to 30 years old! Critical applications software and hardware courses need a leap forward for dealing with today's and tomorrow's world marketplace requirements. Built-in self-correcting design that anticipate internet threats need to be explored and made standard practice for our commercial, military and industrial business sectors.

I own and use both HP and Apple personal computers and iPad & iPhone. All my work up to this book was done using HP and Microsoft Office products. However, not only are virus threats a continual problem, but the complexity involved in getting a report, memo or letter done is becoming so cumbersome and highly complex just to get a document finished in a format one needs to be produced! We are not reaching elegant design levels after almost 20 years of development and so-called enhancements and revisions. Steve Jobs Apple products reflect a degree of simplicity and elegant design that should act as a benchmark for the entire high technology business sector! As we used to say, "keep it simple stupid" and that applies to today's products more than it did back in the day!

One of the top priorities given to me when first arriving at NASA GSFC was to put together a formal Training Course for all the Office of Flight Assurance Management personnel to take. I could not believe after almost a generation of NASA workers and in business since 1958, that no formal training was developed, and a mandatory Space Program 101 was developed for all employees to complete. However, one must remember that this Agency was not military based. As indicated, my military electronics courses at both Fort Monmouth and Redstone was of the highest caliper and even had closed circuit TV and remedial video for those who were having a hard time, one of whom was me! I just had a hard time learning phanastron tube theory with multiple grid flow and spent several nights reviewing the Fort Monmouth US

149

Army TV video instruction...thank you RCA & Philco trainers, two of the best at the time.

Photo 11.8. NASA Space Shuttle with External Tank (ET) on mobile Crawler going to PAD 38 at KSC.

NASA Flight Assurance Management Training Part 1, I called it NASA FAM 101, was given in my first year at GSFC. Our Part I covered the USA Quality history and basics starting from Western Electric Hawthorne Works Era to and including Dr. Deming's New Quality Rules, whose Boston area Quality Seminars, I attended. And from Winter Park, Florida, Phil Crosby's "Quality is Free" added to the offering. I met Phil in Winter Park when I lived there in the 80's. His new, first in the world, Quality College was well attended. I also worked with the same team at Martin on the Pershing Missile as he did when he was at Martin Orlando.

Several senior engineers acted as FAM 101 Instructors (all given Award Certificates, I still have mine!) and we all did our part to make this training successful.

Our NASA Assurance Management Training Program Part I also covered the Fed Gov DOD Quality Control and Assurance field history. As a sidebar to the Boston based Dr. Deming Seminar, I bumped into one of my MBA students, who I remembered from an unforgettable class assignment, when I asked what each student wanted to be in five years hence…she wrote, College President!

And on her Welcome greeting name plate were the words…. Acting College President! How's that for achievement just getting her MBA just a few years back! She commented that my classes were the best…she really enjoyed taking them. I recalled when I started that my classes never would be boring…like so many, especially the ones with PHDs, we all have taken…no sleeping in my classes as I always said that my exams would not cover anything we had not done in class! Note taking was always a consideration if one was to earn a high grade!

Dr. Deming was a world quality pioneer from the 1950s, in both Japan and the USA…he was an icon there and also in the USA, and his sessions conducted in his 90s were not very exciting, but he did add statements like "let's stop burning the toast" and as kids we all remembered that…before today's pop-up toasters, it was common for that to happen! His new rules, however, were about to change the World and he re-invented how work was performed and how management actions should be focused in the workplace.

Putting Dr. Deming's techniques into the workplace was not an easy task. Resistance to change was hard to overcome and cultural changes were difficult, unless like GE, you had someone at the top like Jack Welsh who had the power and position of CEO to make it happen.

I recall a case study on computers in the workplace another difficult workplace evolution. GE at the time amortized the cost of computers across every department so you were faced with the cost whether you used them or not...this was a technique used by many old-school companies to apply the new work tools.

When I got to Martin Baltimore, the place was frozen in the 1950s...more like the 30/40s! Seeing an opportunity to leap forward on my recommendation to bring in the experts...Hewlett Packard... was welcomed, even embraced by new General Management who had just come from NASA GSFC and was the Center Director there! Small world it certainly was. I watched him climb in the management chain 5 levels and in 5 years he became CEO, a job, I found out later, he was hired for from day one. He asked me when on my hiring interview if I knew anything about Inventory control, a little, I replied even though one of my departments at Teledyne was just that. I recall those days that production and inventory control had the highest mortality rate of all the Manager's jobs, so I tried to stay away from that specialty as a Manager.

When I got to GSFC, I asked about the guy I knew at Martin.... he seemed to manage by the book! He laughed and said didn't I know? His wife threw the best parties in Annapolis! Martin sent him to Boston full time to attend the elite MIT Management Program and he was given limo service to work from home for the first year! So much for those Chesapeake boys as they were called...all Duck hunters by the way! One could tell their connection by the mallard decoys on their office credenzas...I still have mine on my home's fireplace mantel...a collector's item!

Baltimore was a tough, blue-collar, hard-working town and working for the Martin Company was a tough assignment as a Manager. One, not a member of the duck club as I called it, needed to watch his back no matter how good a manager one was. I replaced guys who were old timers, some of whom had no college training even at the Director level! This was almost unheard of in the Boston 128 workplace, where even Secretaries were college trained and overqualified to say the least. I wondered from day one why Martin wasn't a billion-dollar company and why it almost was taken over by Bendix in the famous midnight rescue by a Martin friendly white knight called Allied! Martin was then and even today a powerful player in the Aerospace marketplace. I had the upmost respect for Lockheed who now control the reigns. Years later they moved 350 executives to Maryland and took charge of Martins operations...a good deal for all the players. Lockheed, when I worked for them had a great team of professional managers and in my first year there received three in a row raises that made me a very happy employee. There was a big difference in California run companies than the East coast ones...more polished and more professional and finally more fun to work with all around! The only one I truly loved was HP...as they always said, "Have Pride" which we all cherished along with Bill and Dave...both number ones in my book. I always said that HP taught me how to work harder and smarter and to this day they certainly did!

I spent four and one-half years at Martin...just six months short of getting vested in their retirement plan. Despite my many contributions both at Baltimore and Orlando, I never felt at home as I did at HP or even Sanders. Martin was just like many DC organizations, always looking for a fall guy to take the blame for incompetence and poor decision-making. But always there to take the praise for any successes, no matter how small an achievement! Their management succession plan was nonexistent, and the only ones I saw going up the chain were the duck hunters and the Chesapeake Gang, as they were often called. My 401K plans were started here in 1983 and I could purchase stock that for years remained dormant. Only the spinoff of Martin Materials did an opportunity to make some money appeared and after the Lockheed merger it finally took off to over $100 a share. My meager holdings couldn't compare to the millions of shares given to several CEOs since I left. It always amazes me how many million CEOs get, while managers who do their work get meager amounts, even after all their hard work and loyal dedication. Again, to me in hindsight, entrepreneurship is the way to go and without a doubt results in the highest ROI instead of climbing today's even more difficult corporation's ladder of success.

"Computers are for Secretaries, remember that," I was ordered!

These were the first words I was told by my government customer at Goddard. Yes, until 1995, if my teammates or I were found to be using computers we would be fired, I was told at the top of his voice. This I found to be unbelievable! My introduction to computers in the workplace went back to the 1960s. Even then I had in my office, a first-generation ASR-33 Teletypewriter hooked up to the Dartmouth Timesharing System (DTSS) where I could do almost all the engineering research, as we do today on Google, connected to the Lockheed Database and it was all free...the first-generation internet. In the early 70s, I offered to hook up a similar setup (soon to be 12 stations) at Rivier College (now University) in Nashua, NH that got them started in the computer school business. Today, one can obtain a doctor's degree for that same institution in computer science. One of my friends is doing just that as he is teaching full time too!

My previous job at Martin had two computers at my beck and call...one for the company and one for the contract we worked on. Both somehow were charged back to the job! But I always claimed to be ambidextrous...I could perform with and without computers for I had experience doing both. Folks today wouldn't know how to get the job done without their computers as is seen with downtimes...workers are now lost! I needed to again dust off my rusty skills of relying on my personal secretary to get my paperwork done which I did for years.

Photo 11.9. KSC Liftoff of NASA Space Shuttle – truly a magnificent site to witness!

It wasn't until my Civil Service boss retired in 1995, as his Fed Gov 30 years were up, that we immediately acquired dozens of personal computers (PCs) all with the new 286s memory chip along with Word Perfect and Lotus 1, 2, 3! Microsoft was adopted for government use the following year and the first-generation Office was what we all learned at NASA schools on-site. I ended up being an expert with PowerPoint along with Excel and Project Manager. All Office products became a mandatory skill requirement as we approached the Y2K era end. I also loved to use VISIO; a computer graphics tool that could make complex tasks easy to explain with a very professional result. We even got to use Computer Dashboards and the latest powerful management software.

The NASA Space Shuttle named Challenger flew 10 missions, with the final one occurring on January 28, 1986. I was in Florida working on the Pershing Missile Program who at KSC performed several successful night flights from the Cape before the disaster occurred. I recall looking upward from the office parking lot to see the launch that on such a clear day could be seen for miles and miles inland. We all were mourning the Astronauts sudden demise after watching them that morning on TV going to the launch pad, after days of delay due to the downrange inclement weather. That night before at the Cape, I wore my bomber leather jacket for the first time, as the ocean wind was so cold and bitter. We found it hard to believe that the shuttle would launch the following morning.

For the best write-up on the Accident see the Roger's Commission Report on the Space Shuttle Challenger published on June 9, 1986. In that Report's front end is a sketch of the culprit or main cause…a Solid Rocket Booster (SRB) field joint failure. On my first observation, it appeared upside down…like a rain gutter that collects water flow and causes it to flow to downspouts to ground drains. In checking this out comparing military designs I found that they were right-side up causing the water to roll off instead of collecting. The NASA SRB did not use this design and perhaps only a worst-case launch situation would be a cause for concern. However, after analyzing product failures for most of my career it was a high-risk design with cold weather and in rain or freezing sleet. And that morning it was at a very cold Florida day of around 30 to 40 degrees Fahrenheit. Essentially a worse case launch situation environment. History and extensive analysis are covered in the Report initiated by then President Ronald Reagan.

Years later, on February 1, 2003 another Shuttle named Columbia upon return from Space disintegrated on re-entry over the states of Texas and Louisiana on its flight back to the KSC launch site. Again, NASA by August 2003 printed another report called "The Columbia Accident Investigation Board (CAIB) Report. Two books are worth reading, the first where a term called the "Normalization of Deviance" appeared. Written by Diana Vaughn, the book shook the NASA culture. Defined as a "gradual process through which unacceptable practice or standards become acceptable. And as the deviant behavior is repeated without catastrophic results, it becomes the social norm for the organization!" A second book written by Kathleen Fahey, on both the _Challenger and Columbia Disasters_ also makes interesting reading to the interested observer.

Just as a means of comparison, the photo below shows the large Saturn Rocket that I described earlier that was tested daily when I was stationed in Huntsville. It made its way from Huntsville down the Tennessee River on a large barge to the KSC site at Cape Canaveral, Florida. The scale was one of the largest rockets ever designed to take the moon lander on board and return our Astronauts safely back to their home planet...Mother Earth! The Shuttle never went beyond Earth orbit but was a workhorse space plane delivering many satellites and later the Space Station to its permanent orbit around the planet Earth.

Photo 11.10. Saturn Rocket on KSC Launch Pad with Apollo Capsule inside ready to go to the moon!

12. Adversaries Offensive: "Plan to Fail" – Insider Beltway Tactics

I know that there are many stories that you will find both believable and unbelievable here in my book. But my most un-believable one was the inside the beltway "Plan to Fail"! Almost like out of the new TV hit "House of Cards"! Not being a DC insider, and always seeking success in my industry roles, this political tactic was difficult to believe. As a seasoned manager, I thought I saw all there is for inside and outside the trade tactics. It was a new one that I experienced first-hand and survived. This tactic, or strategy if you want to call it that involves placing the target in an untenable situation, with an unsolvable problem. Many successful managers can deal with the controllable factors of the marketplace; it's the uncontrollable ones that usually cause business failure. The purpose was to eliminate the targeted victim, like an NFL Linebacker or linebackers blindsiding an opponent with a triangulation hit as was demonstrated in the days of Joe Montana of the 49ers. It usually resulted in taking the opposing Quarterback out of the game and here the objective is…forever!

Mine came after building and leading my Team successfully for around two years. I recall when studying leadership that one of the big disadvantages of being one was victimization. With the glory of victory, comes the agony of defeat comes from the sports world but it is part of a manager's world also.

After the HST launch from KSC and initial turn-on failure weeks later, it became NASA GSFC's job to get it fixed and fast!!! My eventually 15 people HST Flight Assurance Management Engineering Team assigned to the project had at least 25 years' average experience. Some were supervisors and even managers with leading Aerospace & Defense Department contractors. I needed to quickly add a 24/7 coverage to meet the new demands for Flight Assurance Management personnel…and I did just that forthwith!

The on-board HST Computer DF-224, was made in California years ago, however, it was a weak link or unreliable sub-system perhaps to be repaired or replaced by, back in the day, newer and latest 286-like PC technology. A spare computer was available and bought by NASA. Someone at NASA, who did not like my style and perhaps jealous of my success picked me to do the quick reaction engineering study on the spare DF-224 HST computer. And NASA needed it finished in 30 days or else! They needed to see if it would be worth replacing or designing a new one. This was a setup, my turn to fail, since it seemed an impossible task. The records were still in California and needed to be sent to Goddard. Twice, I was threatened for dismissal, if I failed by both NASA and my then contractor boss. And they were deadly serious too! Ok, I said, show me the photos, better yes, they said, we have one here in a Goddard Center Laboratory.

Their plan for me to fail went this way, I was taken to the Goddard Laboratory where what I call Exhibit A was stored. My immediate contractor boss accompanied me and said well here it is, and you have 30 days to finish the engineering study or else you will be fired. What a back stabber I thought, here was a guy who up to now patted me on the back almost every time he saw me and said great job! Keep it up! Yes, I thought he was what we in the trade called a straight shooter and until now, I had a high respect for his position that was over 30+ years here as a manager at NASA. Now he became an adversary, I was a little upset to say the least. We reached the Laboratory, opened the door and there it was! WOW! I exclaimed, "the famous M60 Tank Fire Control Computer", I replied immediately! Well, his look was my revenge and was one up too! Wide eyed, and extremely puzzled like they picked the wrong project to fail and the wrong guy too…both their mistakes or should I say blunders! I held on and wrapping myself around it, said "I worked on this years ago while Director of a NH Company"! We left with him staggering like a beaten boxer, returning to our cars and waving goodbye in the parking lot. I did not see him for the next 30 days until I completed the study successfully.

I was relieved to know that with the right personnel, we could together pull this off on schedule and without using HST personnel even my team members to help as was the rules, one of the things he told me as we were walking to our cars. I could write a whole chapter in this book on that exercise. First, I need to take my hat off to the Team that helped me succeed. All key players that I had hired and brought to Goddard from all over the USA. The new players worked like the professionals they were, all experts in using the new tools, with computer generated charts, graphs and tables. The legal secretaries finally hired after screening many un-qualified candidates were so valuable and they eventually typed the end Report. It took me several rounds of hiring to find them, all now contractors. And my good friend and colleague who told me almost at the finish that this was impossible, but my early quarterback training told him that we together would succeed. One of my guys later to go high up in CS management rolling over to the government told me later it was the most challenging work he ever did at NASA! And he was so lucky to be part of my plan "not to fail offensive." I never revealed the true nature of this work; all thought it was just another difficult assignment NASA gave to contractors, standard operating procedure (SOP).

I picked two new engineers that I hired from the Boston area to help; both later reached the Director's level in their careers at other contractor companies. One I hired, after I found out he went to Bishop Guertin High School in Nashua, the same school as my son Robert graduated from. He also had a BS in Math from UNH. He became a key player and all the NASA Flight Assurance Managers, Civil Servants, wanted him on their team.

Another top manager at NASA today, helped me do the computer work that other NASA Directorates had, since you will recall I had strict government orders not to use them.

I forgot to mention that of the several secretaries available to do the job, two were part of the plan to fail. My search for a good one paid off with the hiring of a legal assistant who did a bang-up job for our team and for NASA.

After several weekends and burning the midnight oil, I was faced with hundreds of records on my Conference War Room office walls, I created, and that I was given to work the task. Time was getting short now and I still did not know what format and structure could be used to tell the story and make sense of the information we found and the many conclusions and answers to be presented to NASA's top management.

The last weekend before the deadline, the answer came to me while watching a PBS Special on TV. The Special was on the Kennedy's specifically, the Joseph Kennedy Family. Family tree hit me hard, yes, it would work, and I could use that format of connections limb by limb. The following week, we wrapped it up on the 29th day. However, I still was not finished! 24 hours were left, what to do! Here is where my many preparations of college term papers helped. I told my legal assistant to go out and get a big stamp made with the one-word DRAFT on it! Every page of the over 200-page report was stamped with DRAFT even the Executive Summary, all they really needed to read, right up front.

Several weeks later, I was given a NASA Award and check by Astronaut Steven Covey, Captain of the first   post-Challenger Mission Flight and Commander of STS-61 HST Mission that was a big success! And after the highly successful, First Servicing Mission, the $2.5 Billion HST Telescope containing a new 286 computer was on-board and is still is running to this day. After that honor, I felt like President Ronald Reagan…as they used to say in those days…Teflon proof! I never was faced again with a plan to fail, to my knowledge, again and went on to work for NASA another almost 20 years!

NASA Flight Assurance Management (FAM) Training Part II - NASA FAM 102

In developing the advanced FAM Part II Training, I wanted it to be like a college grad school level course. And I also included individual presentations by our Task Leaders, who covered their specific programs and in detail on what purpose the program was and how it fit into NASA's mission. All did an excellent job and I think they liked that part the best from their interest levels.

Such programs as HST, GOES, TDRSS, EOS, TRMM, Space Station, KSC, VAFB, AQUA, AURA, ...etc. gave everyone a good presentation at the state-of-the-art space technologies and spacecraft systems involved.

Part II ended up being an advanced training course for NASA Flight Assurance Managers and NASA program personnel that required Part I completion first. Heavy emphasis was placed on statistics and graphics, as these tools were part of our presentation requirements. The final sessions covered a NASA HQ Case Study exercise that I created and wrote up for teams to solve like the ones I used to use from Harvard University! This one dealt with the current issues that were being discussed at the Fed Gov Agency.

As a side bar here, I recall giving the preparation for doing the statistics instruction to one of my smartest engineers. He surprised even me when handing in his assignment. Almost a 100-page course on using Statistics along with how to and examples with training aids included. I used all he did without touching it and to this day is one of the best write-ups on the subject I have ever seen!

Editorial note: The fellow who prepared that Section went on to become the NASA Chief Engineer at GSFC one of my proudest accomplishments was finding and hiring him! I originally hired him to lead the Space Station Project that was soon to be given to another Center causing me to reassign the entire Assurance team consisting of both hardware and software engineers.

Later, I recall we followed the career paths of our entire Team...over 30 went on to become managers and Directors at many Aerospace companies throughout the USA! So much for proper training of cadre.

Photo 12.1 Below is Certificate of Participation for instructing and preparing the two NASA GSFC Training Programs successfully conducted in my second year at Goddard.

Photo 12.1 US Space Station Module with remote arm positioning it in place on the Space Station.

13. Hubble Space Telescope (HST) Servicing Missions

The several large-scale new NASA Spacecraft Programs were the Earth Observatory System (EOS), Hubble Space Telescope (HST), GOES, TDRSS, XTE, TRMM, etc. Overall, I had 40 bases I had to cover with highly specialized, highly skilled and trained personnel. Since the first-generation engineers were even retiring as contractors after their 30+ years NASA Civil Service careers, it would be a difficult job to fill their shoes. I did have a few long timers working for me and all were a great help in making the transition to a new first-string team.

One good thing was the US Aerospace business and even the DOD side was winding down so many experienced aerospace personnel were looking for new assignments. I was lucky to find several Managers who welcomed the opportunity to come on board.

Let's now see how we got a World Class level of players that got us through the next 10 or 15 years.

EOS (Earth Orbiting Satellites) was a large-scale program with multiple satellites such as AQUA, AURA & TERRA…all focused in studying the home planet Earth. It required a broad range of technology and my guy came from Princeton University and a star player he was on the job. Many players were often management level engineers who kept their hands on the technology, like me…individual contributors who could do both jobs. Some of the new technologies such as today's high storage devices were not even available…. the first-time terabytes of daily storage capacity were being discussed. A far cry from the once 256 bits that some said there was no reason for any more than that! For almost 10 years my Task Leader, Frank Ferraro, led the EOS Team even as a Grand FAM due the scale and complexity of the EOS Project.

Eventually, I found out that my EOS Team leader, Frank Ferraro, a Philadelphia native, was a UASF pilot who fought in the Bay of Pigs invasion and was personally honored by then President Kennedy for his efforts! Also, his cousin was the first female to run for President of the USA! I found this out from his friend who was my Team leader on the HST Project.

HST was next in line.... I searched for a specialist with Optics technology experience and a myriad of other exotic instrument designs. After screening literally hundreds of resumes, I discovered a contractor with almost 50 years' experience. Age discrimination if there was any was in reverse...we wanted seasoned players the more the better. One example was when I flew and that at times seemed always upon entering an airplane I looked for grey haired guys in the cockpit...not to discriminate for one of my best flights was a female captain who gave us a bird's eye tour of Baltimore while waiting for FAA clearance to land...which she did perfectly...Id fly with her many times I told her upon landing. Many lucky flights had a pilot named Scully, the USAir hero on the Hudson. I knew he was a Viet Nam Navy trained pilot so often I got some shuteye even before we took off! My fear of flying if I had any was probably overcome in the 1960s, when I was assigned to a procurement team of engineers who toured all the US microelectronic manufacturers. Day after day I found myself in a Boeing 727 off to the next city on a 14-day tour. Of course, it was before 911 and missing all the delays that are now part of the flying ordeal.

HST work lasted for almost 10-15 years (25 years Anniversary now, as I complete writing my book); it was designed when the industry was thinking modular. All critical functions were part of a works in the drawer design that was accessible and easy to maintain at the time were state of the art, but downstream were costlier than a stand-alone design, like today's desktop and laptop PCs.

Again, on my team was Aaron Pokrass, war hero and part of the greatest Generation, I treated him like an adopted father, and we became good friends. It seems that he was part of General Patton's Army rescuing Bastogne and entering many German death camps. Conditions were so bad he didn't want to even talk about it, but I was so proud to even know him and have him on my team. Still alive today in his middle 90s, I miss our great times together. His dear wife after serving diner one night told me the other side of the story about him and NASA. It seems that NASA a few years ago wanted to hire him and was interviewed by my NASA Directorate bosses. My bell rang remembering when his name was mentioned on whom it was that I picked for the HST lead job. When I said his name, it was like a pin dropped, silent acknowledgement that the right choice was made, but I never knew the entire story until she told me that night after diner.

Prior to the HST launch, that I witnessed from a side road outside the KSC Control Center and had rocket fuel exhaust showering down on us under the bright Florida sunshine, I sent my Team lead to Huntsville to pick up the data package that included test results and evidence of inspection signoff. He returned to GSFC with the following story. When requesting the complete data package, he was told by the Redstone people to take a message to Goddard you will get the paperwork after launch and no sooner! I told my guy to write me a company private memo to that effect and I'll send it to my management. By that time in my career, I had quite a few scars on me and as they say cover your ass or CYA is the name of the game for survival. This memo did just that once HST launched.

On April 24, 1990, HST was launched aboard Space Shuttle Columbia! What a liftoff from KSC Pad 39A …perfect…and to think what was on that shuttle was as large as a Greyhound Scenic Cruiser size bus. WOW! I thought American (and in the back of my mind Von Braun's) technology at its best! What a great time to be alive and working on the most fascinating technology on the planet! Everyone celebrated later at the Officers Club at Patrick Air Force Base quite an ending to a historic day.

Well, on June 21, 1990, as everyone knows today, HST once energized did not work; a spherical aberration was keeping it from focusing on its target. This occurred week after launch and as soon as this was made public, I was called by

168

my company's top management, supposedly, to be held accountable for my HST Flight Assurance Office's Team team's involvement. I took a copy of my HST Team Leaders Huntsville Trip Report and Company Private Memo that of course was by that time forgotten. It seemed that some Corporate VP was concerned that we could be sued for our role in the HST debacle! I presented my evidence and said don't you remember my sending you this Company Private memo? Oh, yes, they replied, I guess you're off the hook. Yes, I replied, we did our job! Now on to the hardest part, getting HST fixed. The Nation's best experts were brought in! I also sent my guy to the mirror's fabrication facility in Connecticut. He called me and told me that the mirror was never tested end to end. No evidence existed. This was despite an on-site NASA Government Resident Team assigned to the facility. At the same time, a VIP Committee chaired by then Senator Al Gore had NASA on the carpet. Why wasn't it tested and why wasn't the competition's mirror bought that was tested...this went on and on for days...I tried to watch it all to see what would result and what would be the outcome. As in most DC committee hearings to this day, except for an embarrassing situation and somewhat of a slapping of the wrist, emphasis was placed on getting HST fixed and fast! My work effort for HST was to come later once the job fix was studied and completed for execution. Not only optics were to be fixed, but the on-board computer also needed updating and maybe replacement too! I think I could write a book on what we found out on Hubble but since people are still alive who would not be very happy, I'll just pass on this assignment! The Hubble Space Telescope Optical Systems Failure Report Nov 1990 can be read at this web-site (ssi.berkley.edu)

The On-Board HST DF-224 Computer Study (copy in my possession) that was assigned for me to get done in 30 days did reveal that the on-board computer replacement could not be used. Here was an over a million-dollar spare part that had a notorious failure record and was worse than the one on-board. It turned out to be the shop queen as we described prototype versions of early designed hardware. Goddard went to work on a 286-chip PC version for the First Servicing Mission. My Teams supported three more HST Servicing Missions on my watch and everyone was successful. An Astronaut named Doctor (MDs & PHDs) Story Musgrave needs to be honored for the work he and his teammates did on orbit to fix HST. Had NASA continued space exploration to the planets, Story was the man to go to MARS and he continued to train for the mission. Hours were spent pulling the drawers of sub-systems out and putting new ones in while working in in space suits was very difficult to perform. It does require the right stuff that dedicated men and women have done over the years with shuttle payloads and satellites on orbit! Later, after his retirement, Story and his son did a great DVD on the Space Program and it is available at the many NASA Gift stores (look on-line)...quite a wonderful presentation to a USA Space Hero!

Another long term GSFC managed project was the Tracking and Data Relay Satellite during my tenure being built at TRW in El Segundo, CA. After the Apollo Moon landing, there was a challenge to reduce the worldwide need for tracking stations by engineering on orbit satellites to replace the ground tracking network with a newer one. What resulted was a new on-orbit tracking network called TDRS. TDRS-A was first launched from KSC in 1983, on STS-6 Challenger. It ended life in 2009! My Team covered seven consecutive successful launches of this technical marvel that took place using the Shuttles: Atlantis, Endeavor and Discovery all from KSC. The first one occurred after the fatal STS-51L Space Shuttle Challenger Disaster on January 28, 1986 on which TDRS-2 was lost. NASA had a resident Goddard Team stationed at TRW and my Flight Assurance Management team members were part of them. TRW was one of the powerhouses in the Space business going back to its original founders: Thomson, Ramos & Wooldridge, all space pioneers going back to the 1950s. I was always impressed by their attention to detail, a meticulous endeavor required by such a very risky business as was the Space Program. The responsibility of my team was to oversee operations and act as a second pair of eyes for NASA to witness critical processes and assure that the spacecraft was designed built and tested and delivered to KSC meeting all NASA Standards and Requirements. From TRW, the team support then shifted to my KSC based Goddard players, who received the handoffs seamlessly. Both TRW based, Goddard based and KSC on site teams did that job without an incident for many years thanks to that excellent TDRS Project team. My monthly visits help build a close working relationship between all team members and I was very proud of the work they all accomplished especially after each successful KSC launch.

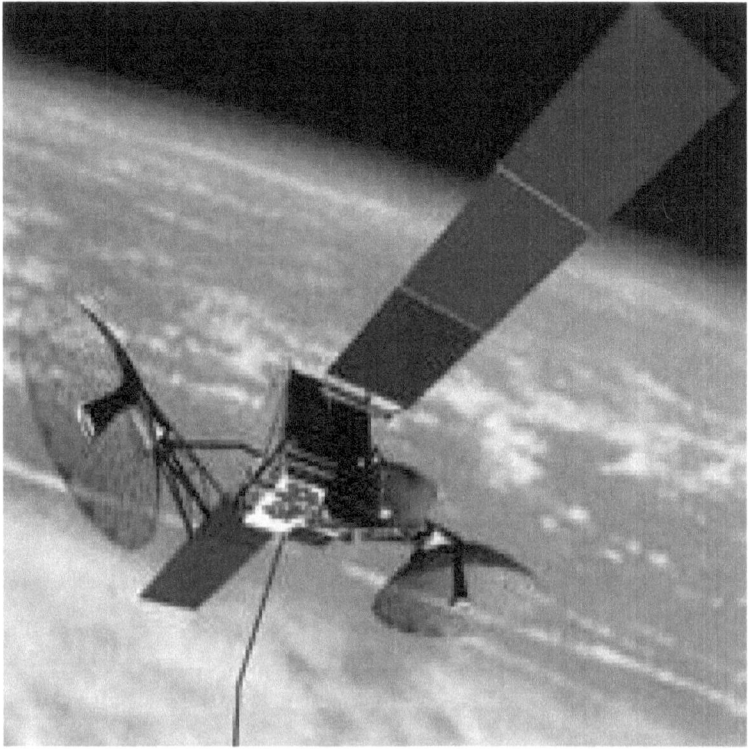

Photo 13.0: NASA TDRS Satellite on orbit an over 30-year endeavor!

KSC is a World Class Operation and precious American and International resource. I watched many missions with teams from all over the World. The Germans who built spacecraft with a wind tunnel look streamlined in every way to the Japanese who micromanaged all operations with meticulous care that they are famous for. Our GSFC on-site crew were experienced contributors going back to the original launch of the going back to the 1960 and culminating in the first test shot in 1981 and operational flight in 1982. Many were ex-military veterans who cut their teeth as I did on military rockets and radar tracking systems. I can say that a brotherhood culture existed among all the players at both KSC and all the NASA operations. Each respected the others job and operated as a winning team in all mission work endeavors.

Photo 13.1: KSC Florida Vertical Assembly Building – showing two paths to Launch PADs, 38A &B

The VAB is the world's largest air-conditioned building on the planet. I was amazed as to its size, where in addition to housing the Space Shuttle on top of the External Tank houses the Command & Center for conducting all KSC launch operations.

Seeing the process unfold when the Shuttle is transported from the VAB to PAD 38 A or B is a sight to behold! American engineering is displayed at the highest levels. Tons of rocketry are carefully choreographed to the launch pad with the grace of a ballet dancer. The entire prelaunch show is one of our finest achievements of a World Class Operation and going on for almost several decades of missions since the 1969 Apollo Moon mission. A one of a kind, USA resource that is needed to be enhanced and nurtured until higher levels of technology are achieved by the next generation of rocket scientists!

Photo 13.2: NASA 25th Anniversary Commendation for HST Servicing Mission (2016)

14. NASA GSFC GOES Program (goes.gsfc.nasa.gov)

Geostationary Operational Environmental Satellite (GOES) since October 1975, was another on-going mission at GSFC. GOES R will be launched in October 2016 **(see Web-site GOES-R.gov)**. My Team Leader, Luke Deemer, was a Boy Scout Troop Leader on the side and an outstanding contributor over the years both off site as a resident in a nearby Aerospace facility and at Goddard. To my knowledge, he still is working assignments for NASA. When I selected him for the Team lead assignment my Contractor boss said, "but he doesn't have a degree!" I knew many major players who never finished college for some reason or another. Many first-generation high-tech workers did not have degrees but did their job no different than someone who had one.... sure, they could not write a book or handle high-level equations, but they had the right stuff too! GOES eventually required both hardware and software support and my software counterpart did a great job of getting the right players for that job too!

Photo 14.0 Kennedy Space Center Visitors pass for AC-77 GOES J Launch. Note all the companies who had major sub-systems on the big bird!

We did many International joint venture launches. I recall two in Russia where on the first one, everyone got very sick on the food and water. On the second, they brought their own food and stayed in military facilities that they said were better than the civilian one the first time. My Team Lead, Gene Volpe, there was more of a super tech kind of guy. A hands-on type that fit into the launch team for his, "let's get it done" attitude. His government counterpart always pleaded with me not to reassign him for he was to valuable a member to replace.

Upon returning from his second trip to Baikonur, Russia (their Cape Canaveral) I asked Gene what it was like over there. What he told me was quite startling! He replied Chuck; you remember the cold war and how we were in competition with the Russians. Of course, I replied, well you know when we go to the Cape and prepare for weeks for a launch? Yes, I said! Well, the week of the launch, in fact for 24 hours we waited and waited for everyone and then almost hours before the Russians brought out this mobile rocket and in minutes it angled upward and blasted out of sight.... then, he said he saw first-hand, what we were facing during the cold war and missile race. As a Russian General once told his American counterpart, you Americans are always worried about quality, we Russians think there is a certain quality in quantity…you hit us with one shot we hit you with so many! Let's thank God that this is in the past…we are now using Russian spacecraft and their engines for many Space applications including bringing our Space Station people to and from space!

The second big job we had overseas was in Takashima Space Port in Japan...a little island off the coast of Japan. On this assignment, two of my best hourly paid Technicians were required to accompany the spacecraft on an USAF C5A and take it to the launch pad. Both had worked for years on similar assignments with no problems. Once travel orders were processed, I had a visit by our new boss who asked if I knew what I was doing! Both people were black, he said. And, I replied, not to worry both were highly qualified and I trusted them both to do a good job. His concern was of course unfounded, and the launch came off without a hitch. I was glad they got paid for every hour in the air, port to port. Few knew, like I, that our hourly paid folks got paid very well for their long hours on the job and deservedly so.

Initially, one assignment that I did prepare a great team for that disappeared to Houston and the Johnson Space Center was the International Space Station. I had a Team of several hardware and software personnel with another Leader who was there for several years. This Team had a Gov'y, who later became my advisory because one of my guys had a behavioral problem difficult to see and correct in his eyes. It seemed this guy was suffering from a serious brain disorder that eventually took him out, but not until it was too late. My only suspicion was he always wore headphones at his desk and seemed to avoid discussion. The Gov't wanted him fired and blamed me for not doing it. Only later we got together and resolved the conflict, but after he created a lot of damage along the way. The game inside the beltway is no different than outside but it does have a special twist and like the poor husband who is last to know about what is going on…the victim here is also last to know too!

The Government guy went on to become one of our highest-ranking customers and he seem to hate my company, our management team and of course me! Later, due to his behavior (I call it the looking glass complex, they are what they see) he got reassigned to a job he knew nothing about and created more havoc in that arena. The Beltway is loaded with players like this…always adversarial, always demanding attention, tearing down instead of building up organizations in my eyes, just for the hell of it! Again, thank God, they never get to the highest levels of command. Although in recent times this seems like a current trend.

Here come the computers ought to be a paragraph title in this book. Always one to use the best tools and techniques used since my Babson College education and graduate training. I could not believe how quickly computerization took place in our Directorate at GSFC. Perhaps it was that all of us new people came from companies that used them for years and were part of our toolboxes like a hammer is to the carpenter. Soon timecards, personnel records, reports, project databases were part of everyone's workplace. This did make management's job easier...however taking away the Secretaries and if you were lucky having an administrative assistant to help the new paperless system, separated the big wheels from the many lower spokes.

The next big NASA Survival Challenge.... Fixing the orbiting Hubble Space Telescope (HST) – Do whatever it takes!

So serious was the on orbit HST failure that one day I was called into my counterpart's office for a discussion on what to do. He began by telling me that NASA was in deep trouble and we needed to make sure that whatever work was done needed to be scrutinized and monitored 24/7 until launch day when all repairs would be done on orbit with space walks by hand-picked astronauts. I was to do whatever it takes to cover all the bases on a 24/7 basis. This meant finding additional people to immediately man the stations from Goddard right down to KSC. Still active in my Baltimore trade circles, I found out that some of my former guys at Martin were retired and still available. This was one good source of experienced personnel. As for KSC, I ran an Orlando newspaper ad indicating "KSC Open House" at the Cape for potential Space personnel. What I got from that ad were former KSC retirees still willing to work for NASA. After hiring a few, I was told "good job" by several KSC Managers who knew all of them. Goddard HST people worked 10-hour days and more and 7-day workweeks too for the duration even after the fixes were made until HST Certification was completed. One must remember that at John Hopkins University in Baltimore, a team of scientists were waiting to conduct HST experiments for months after initial launch in 1989.

Daily reporting began this time early every morning. All operations involving the new fixes were reviewed over and over. Weekly Status meetings covered all areas of concern, the old NASA spirit flourished to make sure the fix it job was done right this time! In short what was done was to added corrective optics like glasses on people to the original mirror that was out of focus. A complete Report is available from the website NASA.gov for those who are interested.

Finally, HST Servicing Mission Number 1 was ready for implementation. The Team all was sent to KSC for final launch preparations.... everything was double and triple checked. Successful launch occurred on the shuttle loaded with payloads of fixes and even improvements such as the new on-board computer. I found it amazing what could be done by NASA when the chips were down. Like a sleeping giant, it was able to overcome all odds to attain mission success.... as shortly occurred in space. Another, "failure is not an option," goal was achieved by many heroes...with the Astronauts again the biggest heroes of all!

HST went on for many years yielding photos and universe maps like no other means. Scientists were given a new tool to explain how we were formed and what is out there for the very first time. One conclusion that I made is that we humans are the only ones out there for HST looked light years out and nothing seemed alive in the large universe of space...but that is only my opinion, I'm hope I'm wrong about that observation.

At one point in management at GSFC, I was going on almost 100 direct reports, too much for anyone to handle. From my college training the magic answer for direct reports is 12 people, like it was over 2000 years ago. Studies were made over and over always coming up with that same number, as the optimum. So, I needed help and knew whom to get from my Boston years, a former high technology Manager who reported to me and was a friend from my first assignment in Nashua in the mid-1960s. I brought down Gus Bonenfant from New Hampshire and told him to not drink the water and be careful where you reside. He followed my advice on the former, but after a TDY job at the University of California at San Diego his nearby residence was broken into and he lost all his new homes furnishings. So much for the Heavenly City's welcome wagon greeting!

Earth Observation Sciences (EOS) is a big part of NASA Missions. After all the new EOS Projects such as TERRA, AQUA, & AURA, I felt comfortable with these sciences since perhaps as a student at Northeastern University, one of my electives was a yearlong course in the Earth Sciences. Not knowing that in my career I would be working on the largest investigation experiments on that very subject. Comfortable, yes, because I excelled in the course earning an A every semester. But this work was not as easy as a college level course and it was for real for any blunders or errors could involve one's life even my own! NASA makes the job seem quite simple but that is only because in Vince Lombardi's famous coaching words "practice makes perfect!"

In my first ten years at NASA, we supported over 100 Space Missions.... domestic and International. NASA partnered with many USA Universities and top Aerospace Industry leaders a hard combination to beat.

XTE (now called RXTE) was a brand-new job that had many new players not NASA old school veterans. The new Govies often came from nearby DOD plants in the DC area. As job openings at GSFC opened due to a retirement or promotion any gov't civil servant could bid for the opening even if they had no NASA experience. As the first generation retired mostly after a 50-year career (30 and out in gov't and 20 as a contractor) outside the beltway career people jumped at a choice assignment. Goddard had the highest paid jobs of all nearby agencies.... reflecting in the rigorous engineering training necessary to qualify for them. One Year Calculus provided a high hurdle for entry prior to the latest only courses in computer science was a requirement to obtain an engineering title. As in any attractive job situation even these rules did not apply if one was a connected to the insider good old boy network...soon to be a good old boy and girl network. I had one "client"...love that beltway term, in place of the customer, as I was indoctrinated to define them.

She only wanted fair-haired young engineers…no seasoned veterans for her. I obliged by asking her "who do you want…give me a name! I did my best to accommodate her, since she wielded a clout with top management.

Back to XTE…. after meeting with our clients two new ex-DOD plant representatives, I knew that a manager was what they needed despite their deer in the headlights look when they tried to describe the person. In my book of screened applicants was a senior IBM manager retired who still wanted some action. A native Baltimorean, with a Johns Hopkins engineering degree thanks to Korea's GI Bill. He was just the person for the job…years of hands on experience and savvy in the Chesapeake mindset of ducks and social interaction. His war stories paved the way for him with all the Project managers and worker bees as we were all sometimes called.

I once asked him after becoming good friends, how he ever got through Hopkins with an engineering degree and he said that math was his most difficult subject and he was offered a passing grade if only he promised to never take another class with his math Professor Instructor again was his answer. He went on to do a hell of a job for a new GSFC Manager.

One example was I was called to the shop floor one day to look at a concern of his. We went over to the spacecraft and he showed me a mirror-like covering that was used to cover certain areas. We were all familiar with this cover it was used on every job on the floor assembly area. In fact, it looked very scientific to say the least. Well he checked the part number of that material and found it out to be highly conductive. A NO! NO! In its application. He did his homework contacted the supplier and they admitted wrong material was delivered. No one knows how long or how this foreign material got to be used, but it did give me ammo to get an Incoming Inspection function at GSFC that is in operation to this day checking all spacecraft parts and materials for compliance before accepting it to stock. Something the DOD required since the days of Korea.

The second big find he came up with was found on a Saturday morning as I was awakened at home. Come down please he asked me I stopped the job! Well, in those early days, stopping a job at NASA almost required an act of Congress in DC. He said something was serious and he needed backing. I got to the GSFC Shipping dock and the spacecraft was loaded for shipment to the Cape (KSC). However, he discovered the wrong truck was sent...no humidity or temperature monitors were in place as the shipping paperwork indicated. A second truck came.... same problem...the trucker said that these trucks were used for years to escort spacecraft to KSC without a hitch! Finally, trades show truck appeared days later, and the spacecraft was safely transported to KSC successfully. So much for making sure things were right, afterward, some managers were complaining that the shipment was stopped...the same mentality that was later uncovered as a deviance culture and I agreed...do it right the first time. Many selfish awards are given out in DC (praising the non-participants they call it) and I thought my guy should have received one for a job "well done"!

My Team stopped many jobs during my tenure, because they knew I would back them up and expected them to do it if required. On the HST Servicing mission we had the power to question anything that seemed questionable and many a times the problem was corrected before moving on to the next station.

Here come the computers...our new tools and ever-changing work environment. The then use drafting and design techniques evolved to computer era tools in one fell swoop. As I said earlier as a seasoned manager and computer user since the 1960s, I was shocked with not being able to use the power of the computer in my daily task at NASA Goddard. As a Boston grad school MBA, I was very lucky to use both HP and the computers in all my courses at Babson College. My Babson Class of 1973 was the first class to use computers in every course we were required to take from Matrix Algebra to Management Information Systems, all

181

contained a computer systems application. Babson, at the time, and still is today, the number one rank school in the world for its emphasis on entrepreneurship. In all my earlier management jobs, I was fortunate to apply computer technology in inventory control, quality assurance control, and factory operations. And even started a new Computer Program and Laboratory at a local university in New Hampshire where today, they even offer PhD classes on the subject.

Once our Civil Servant Manager, who was anti-computer retired, our PC floodgates opened, and I was fortunate to ratchet rapidly catch-up by taking every computer training course the government had to offer. Soon I was using Excel, Microsoft Office PowerPoint, and Microsoft Project Management in all my work assignments. And once I discovered Visio, it too became a powerful communications tool. My last NASA presentation is available on the web a PowerPoint presentation I'm very proud of I became an advocate of computer applications and as a Senior Systems Engineer once I returned to the individual contributor status, I got into dashboards I and INCOSE Systems applications. One of the most powerful tools for my money was the use of computer diagrams. Once understood, these proved to be invaluable in communicating complex relationships and subsystem and system operations. The keep it simple approach was a fundamental role in using any of its techniques we implemented. Scoreboards were also developed for management reviews. Project Managers, using this practice are held accountable for their progress by just looking at critical variables such as materials usage testing and inspection results subsystem status and overall measurement of planned versus actual variances.

Weekly Reports were part of our contract deliverables. Each Monday NASAs top management reviews the past week's progress, project by project; with PowerPoint presentations being made on all my team leaders assigned to various NASA projects. This provided a good method of accomplishment and accountability as slips in schedule and problem issues could be addressed and closely followed. DOD management does similar reviews with Quarterlies and Critical Milestone tasks are dealt with more formality and pomp and circumstances as daily and weekly meeting reviews.

As once said by US Naval Aviator, a "heads-up" display of events is the best vision for determining status as exists today in measuring performance. I recall during my Martin days, a Rockwell B1-B Test Pilot's response when asked what his biggest fear was in test flying…. **"foreign objects!"** he quickly replied! It seems that several root causes for failures in aviation were either adjustable wrenches or soda cans or other debris lost deep inside the aircraft. This was also a major concern in building NASA spacecraft.

Photo 14.1: Heads-up display gives pilot an instant analysis of flight instrument status without needing to look down

For example, at the Cape, before going up to the launch tower Pad 32 A or B, one needed to place their identification badge in a ground floor hut and use another badge and tether their glasses. In addition, one needed to remove all jewelry prior to visiting the shuttle. One funny story was on one of my double check work trips to Pad 32A, I got on the elevator with one of my newly hired engineers. As we approach the top, over 300 feet above the ground, in a cage elevator that exposed all sides, and grated exposed floor below, my new guy panicked and held onto the cage with dear life. Later, I found out he was terrified of heights, but did not want to tell me! We teased him until he finally got over his fears after many successful trips. I recall the first time, I went up and saw the astronaut escape slide to the ground, a small basket that was tied to a tight cable from the top of the gantry to a flexible like ground protection that was an escape route if needed. I doubt if it was ever used, however, I wondered who was this contracted to and how was it ever tested. It was a daredevil task if I ever saw one.

Photo 14.2: NASA Space Shuttle Ready for launch at Pad 39A

I recall using the KSC Headquarters restroom one time in the mid-1990s. Almost all the sink faucets were rusted and needed a replacement, as no money was available at this time for fixing an almost forty-year infrastructure! All I could think about was the foreign teams, German and Japanese engineers that must've thought about good old USA from seeing such an unfit facility. Salt air sure does a job on all metal surfaces. I recall doing three-day salt spray tests at Sanders and being shocked of the damage done to electrical connectors.

I also recall a new upscale condo going up in nearby Cape Canaveral and thinking that I should buy one as an investment. Little did I know, visiting it as a hotel years later, that all the doors needed replacement as their hinges decayed and most of the roofs leaked due to the harsh salt environment. I wondered also, how the people did who bought them, with long-term mortgages, feel after 10 years, when they would need to find a new home or completely re-build their existing one. So much for great PR job those Florida real estate people do! Caveat emptor should be in every major Florida mortgage contract.

15. Beltway Government Contractor – Third Line of Defense!

As a government contractor, let me clarify my role for people who do not know how our, of the people, for the people and by the people, form of government runs. As I indicated earlier, redundancy was built into the political system, just like it was built into all our planning to at least the third backup level. Start with Plan A, if that fails go to Plan B and finally Plan C or D in special cases where criticality may be involved like loss of life!

Here is how it really works! Even before my DC inside the beltway days, my initial perception of the bureaucracy was as most Americans, to say the least, naive or elementary. First, there are three levels of government: and the answer is not the Executive, Congressional and Judicial as we all learned in grade school. As anyone like myself who is skilled in Organizational Management, I would like to describe for you, probably for your first time, all the levels involved. The first levels are those high-level elected people, officials sometimes referred to as whom we or someone picks for us to elect to an official governmental office. They include appointees, like Jackson said so eloquently "to the victors, belong the spoils". These folks, I classify as "Govies"! These top job slots occupy the highest levels of government and pay grades too often-times reaching over $200,000 plus expenses a year! In addition, elected US Congress persons and US Senators often get the same base annual pay for the rest of their lives after just a few years' work. Not to mention, in addition, the best paid medical care in the World and most vacation and holiday pay plus personal time off than any other levels.

The Civil Service structure consisted of GS 1 to GS 18 (for example, Rocket Scientist Dr. Werner von Braun was placed at that highest grade). A military-like ladder organization then exists with lowest level recruits, then privates' right up to up to the highest level called Generals and Admirals; and as they say in government and top management too, rank has its privileges (RHIP) in force at all grades! From offices furniture and size to even computer screen sizes, the pecking order existed everywhere! All wanted, it appeared to me, to use a God Father Movie term, to "wet their beaks!" We contractors, the third level, for example put in as a minimum 1,980 work hours per year. But not the Go vies, at least the ones I knew! Subtract a typical official 12 national holidays off, plus 240 hours of vacation and 30 days' personal time (another 240 hours) and the net time worked is almost 600 hours less than the 40 hour/week bottom feeder jobs as some are called by the top-level sharks! Multiply that 1,980 (248 days) number (net 1404 – 176 days) by the hourly rate of $50 per hour and you are reaching the level of USA Doctor's (MD) pay or $100,000/per year! Now divide $100,000 by 1404 and you get $71.22/hour or $148,148 for actual on duty pay rate! How would you like 72 days paid days off (576 hours) instead of that 80-hour per year break you earn in a civilian job?

This level of time worked also applies to most of the next level of who I will call will loosely call "worker bees". This next group are a second layer of people called "Civil Servants" (CS)…also referred by me and others as the real Govies because unlike the first, except for a few at the highest level, these jobs are in fact life-time career employees. In addition, you have the Senior Executive Service (SES) at every branch, including NASA, appointed political officials put there by the Executive Branch.

As I said earlier, I myself had a ground floor offer to become a CS GS-9 (around $10K/year in 1961), when I left the Army in Huntsville to join the new Agency called NASA. However, I considered it a low paying job when I could make as much money in industry…so I turned it down. Little did I know that in 30 years at 50, I would be retired like friends who did sign on and received great pensions almost like a CEO in industry? The Third level was of course my layer of civilian workforce, the so-called worker bees or bottom feeders in today's terms, Contractors! As President Harry Truman (No college grad here) once said, **"The Buck Stops Here"** and it did on those few winning

Contractors…like Lockheed, Martin Marietta (now called LOCKMART), your number one weapons store and of course Northrup Grumman! "Thank you! To the Long Islander guys for many of our NASA mission successes, we could not have done it without you" was said by the famous Apollo 13 Mission Control Chief Gene Krantz said after a photo session at GSFC. He was the one who coined "Failure is Not an Option" in those NASA pre-Goldin (another story about a new NASA HQ Administrator) …actually Goldin years! We all loved and respected Gene, he represented our USA and American values at their best! Upon retiring, he contributed all his collectibles to his local Texas Church group and accepted no payments for all of them.

Hear…drifting again back to us NASA and DC Contractors. Here is where the rubber met the road. Again, in today's terms, we were the worker bees. The Contractors really do most of the grunt work in our government and at the best Agency called NASA! Our organizations had a hierarchy too. A ladder type reporting structure upwards or downwards wherever you were. From entry level to Supervisor to Manager and Director finally up to VP, Sr. VP and Executive VP to CEO. As I climbed this ladder over the years up and down, I found out something very interesting…the higher you went the less you did and the higher you got paid a form of Pareto's Principle…. the trivial many are managed by a selective few. It seems grossly unfair but as President Reagan did (voted many times by Fortune as the World's best Manager) delegate and hire the best players to be on your team. Lead, follow or get out of the way was a mantra by those I watched, who pushed their way up the ladder.

As a DC Beltway outsider most of my career, I did find it appalling to discover so many what I called redundant positions especially in Management. An old boss CEO told me one time that having a two person on one supervision level was a no to him and he was right. He never would have accepted several supervisions on ones, as I was the case everywhere, I saw in our government. And at NASA, voted as the best of the best-run agencies in Fed Gov't, it also revealed many a no no's in the organizational structure too! Deputy dogs as we called them were everywhere in government and industry program management structures.

My position at NASA was called a working Engineering Section Manager and I liked it that way. I was able to make engineering contributions, learn the latest technology…Flight Assurance Management of large-scale systems design and analysis was my game and I loved it…as many of my Team mentioned many times. We are here to earn a living they often said, you are here because you love it. And I certainly did, all 50 years of it! High technology is like an addiction and many of us had it running in our bloodstream. The more difficult the task, the more I was challenged to get it done! I was a hardball player, a disciplinarian, but always fair and just. And not a back stabber, as many players inside the beltway were…their own key to survival. I always hired the best, more so, if I felt they were smarter than me. It paid off over and over for me, as few times, I needed to micro-manage like so many other disliked adversaries did. Hell, most of my team was ex-managers who still could play the game and get the job done, on time and in budget…and most importantly with a high level of World-class quality excellence!

As an active NASA Manager during two government shutdown periods, I was amazed how efficient and smooth working everything became. Both worker bee efficiencies and effectiveness seemed greatly improved during these periods, where we contractors remained on the job while all Govies were at home relaxing. I did get upset after finding out that their Union…. yes, they have a Union, fought for and won back pay for all the time they were off. Yes, you can't beat a CS career with Uncle Sam!!

In 1989, the NASA setting was as follows. Approaching their 30th year anniversary, many first-generation players did their 30 years and out and either retired to the sunshine state or remained on board as contractors some in the very jobs they retired form! They were called "double dippers" and were proud of it! Little did I realize that I too would join this exclusive club after my second and third time retiring?

During my first week as a new Manager on board, I was taken around the Center and introduced to all the key players in charge. Introduced as that QA Manager guy from Martin Baltimore, I was immediately given the good old boy treatment with their own stories of the infamous or famous place in Charm City where over 65,000 people worked on WWII bombers around the clock. I had no found memories of the Chesapeake boys' treatment of me even when I took the factory from the 1950s, WE Hawthorne works culture to the 1990s computer culture in my short tenure there.

My Government CS boss told me my new mission. I want you to find me and hire the best next generation of players…I don't care where you get them or how much you pay them, only get them fast and get the very best! What he had facing him was a backlog of multi-million-dollar NASA Satellites called GOES, TDRS, Landsat and HST to name a few big ones, resulting from the Space Shuttle Challenger disaster event. And a diminishing workforce with big manpower holes to make sure things were done right. I found out this was NASA secret for success. It was not with the entire "how to" documentation that collected dust on the shelves of their offices. It was the experienced, self-starting, down the learning curve workforce, similar to the mindset of those WWII soldiers with American bred values that got us to be first to the moon and back…first thought of as impossible in my lifetime…certainly ridiculous in my Father generation too!

My first Management Mission task: Hire the **best** for NASA GSFC…and we screened thousands to get a few hundred candidates…looking for the "right stuff!"…experienced and highly skilled USA Aerospace industry contributors.

This mission took more than me to pull off! I had many other of my Team help me do my job. And I had one of the best HR guys, named Cliff Stanley, around to do the first screening of literally thousands of resumes from all over the

World seeking employment with what I call today the best of the best, "Top Gun Space Experts" in their specialized fields. Champions, like the NFL do not happen by accident. It takes a work culture and yes money to get the best there is, but there is more to it than that. Let me walk you through the process.

Cliff one day told me the following story. "Chuck, do you know how difficult it is to get a job here?" "Ok, tell me," I replied. Well, it's about the same chance as getting into Annapolis…the Naval Academy just down the road from here. Perhaps, he continued it is even more difficult; one in 1500 applicants are hired! Quite a statistic, compiled by someone I considered a seasoned pro in his chosen field. For the next several years my job, using all my skills, training and experience focused on planning, organizing, staffing and directing a USA based work force whose responsibility was making sure NASA products had the right quality, all the time without exception, to a job with results better than six sigma level. We had over 15 NASA USA based worksites and they included both launch sites at KSC and VAFB to accomplish our missions.

One of my proudest days was when my HST Flight Assurance Team received 36 NASA Medals and Certificates of Honor HQ Awards, including me, for a job well done on our exemplary support in fixing the Hubble Space Telescope, called the First Servicing Mission. It took almost two years of effort to prepare for on-orbit repairs to be made on a new Huntsville designed and built telescope. As the press so often displayed, a flawed optics mirror that was launched untested was a formable task to repair.

HST never had what in the business is now called full-up end-to-end testing. In my DOD days, called Final Acceptance Testing with Government oversight was a contract requirement. And with the David Packard Commission Report requiring In-Plant Reliability Acceptance Testing (Fly before Buy, we called it) I found it hard to believe this could happen at NASA! It seemed my old Army buddies in Huntsville really dropped the ball on this job! I recall sending my most senior Engineer to MSFC prior to launch time to bring back all the documentation since GSFC was now given the HST mission after launch. He came back saying the head Gov'y told me to take a message back to Goddard, "When it is in orbit you will get the documentation!" Sounds like another Baltimorean & DC refrain saying, "when we pass it you will know what's in it!" Or words to that effect!   We will cover more of HST later for it makes a marvelous story of courage and perseverance by the Astronauts and unheard-of contractors that ultimately resulted in overwhelming mission success.

I used my classic line and staff textbook management & organization architecture to re-organize the Flight Assurance Office managerial organization chart. The CS head loved it as he was at the very top, mirrored by my staff in every position. We were the ghost organization behind the official Fed Gov't organization was my managerial tactic.

Immediately, I went to work with laying out a new organization chart. One of my players had a pc with a suite of IBM software like OFFICE (he needed to do some work for the Gov'y guy who told me not to use them…with one exception of course for his CS work!) He retired from IBM and was one of their Director's, who was well known in the business, and who I even knew about, his personal assistant.

We collaborated and drew up several charts for our Gov'y boss to bless. Of course, he chose the one with him on top. I had my guy mirror his chain of command with me on top! He liked the idea…I did the work and he took the credit…typical DC Beltway practices!   I recall one of my Boston Professors in my Northeastern Human Relations Management Course telling the class that unless you are on an organization chart your nothing and I guess from the business world's point of view he was right! What good is a scorecard without a line up, we used to say as kids relating to the sports arena.

We finally resulted in a business-like structure from the Director level on down to the lowest engineer who reported to what we entitled Task Leaders (short for Supervisors) centered on major projects they were assigned to. This was right out of a Management & Organization 101 textbook not part of an engineering curriculum, therefore not practiced well here. I taught the subject years earlier at the undergraduate level and was able to apply the technique quite readily by job function and work specialty.

16. Management by Walking Around (MBWA)…Round Robin to NASA Suppliers & USA Launch Sites

The first year at NASA Goddard, I spent most of my time in the air…first to Fort Wayne, Indiana…what's in Fort Wayne and why, I first said knowing the remoteness. This was answered on my first visit to ITT, eventually a fabrication house that we brought up to NASA and Aerospace Standards. After walking around the plant, a large DOD supplier, I spent time in the General Managers office and gave him some bad news! I used the same method the FAA guys used on me at Kollsman in NH, when I was a Director there. Is this the way you want to run an Aerospace factory…and let him react? Why, what is wrong he replied. Your facility is not up to NASA Standards…no clean rooms; no power backups, no segregated work areas, no training program or NASA certified workers I started with.

Oh! He said, we need to fix that. I told him we were going to help him by putting in a team of experts to quickly correct the situation. He was damn glad to hear that and thanked me for my help. I spent many days myself in Fort Wayne and had 'Get Well" teams of players there for months TDY. Oh, I forgot to mention I did on my first visit also found out why NASA work was in this remote place. Well, who's the Vice President from Indiana…was the obvious answer! He was also, at the time, the highest level on the DC government organization chart with the NASA Administrator reporting direct to him!

Next stop, my old neighborhood, Palo Alto, California…new home of now what was called Silicon Valley! When I worked here in the 1960s there were around 30 companies who populated the region…HP and Lockheed, Xerox, Varian, Ford and the whole new micro-electronics industry family…Intel, Fairchild, National Semiconductor, etc. and of course NASA Ames Research to Moffett Field nearby. Almost 30 years later, 3000 companies now called this area home…and growing rapidly. Instead of flying into San Francisco airport miles away, we now went directly to San Jose Airport. Gone were those beautiful farms with the freshest of fruit and vegetables, I acquired on the way home from work when at HP. And, no more quick rides down El Camino Real to Mountain View for lunch, traffic was becoming as bad as the DC area where I now resided. And Apple was on the ropes, suffering from new management that did not know what they were doing; gone was boy wonder Steve Jobs, replaced by whom we all called the Pepsi soda guy. I looked at buying some stock (good buy at $1 & 7/8 a share) …and even toyed with going to work there to help straighten them out. But chaos keeps me away, for who knew they would survive and become even bigger than HP or IBM for that matter the biggest on the planet. Google also was a new neighbor, but a guy named Larry Ellison, (another college dropout with $56B now!) had a tiger by the tail with his new software company called Oracle!

Photo 16.0: Our favorite San Diego scene…Hotel Coronado, not that we could ever afford to stay there… "Some Like It Hot" movie filmed here in 1950s…everyone thought it was filmed in Florida!

Ford Aerospace (now Loral) was also there from the 1960s on San Antonio Road, a short distance from where I lived in Mountain View. It was now called Loral Space Systems, bought by a NYC firm I recalled from the 1960, when a group of ex-employees became part of my group at Sanders Associates in Long Island. It seemed a guy named Bernie Schwartz loved to play monopoly and acquired many classic companies along his way to fame. NASA was one of his best customers and I liked the team of players there. A NASA CS Chief named Billy Marshall became good friends with me and my guys loved working for him at that location. Later, I will discuss what happened after he and my team of players left watching the store and retired…. a big event and embarrassment for both Loral and NASA that could have been prevented.

From San Francisco, a place that I always loved to visit and live, it was on to Los Angeles, a new workplace that I placed a little above Frisco after seeing just what it had to offer. And LA certainly does have a lot…from neighboring San Diego down to Ensenada, Mexico, one attraction after another, still finding new places since my first visit in starting 1967. The Getty, Beverly Hills, Wilshire Boulevard, Semi Valley, Yorba Lynda, Tar Pits of Labra and of course VAFB and Santa Barbara were often places I visited many times.

My off-duty home was the Marriott at Marina Del Ray where from the late 60s, I watched many Super Bowls in hotels there escaping the harsh New England winters even for a few days in the months of January/February. Later, when my NASA Colleague Gus Bonenfant was our NASA Resident at the University of California San Diego Campus Laboratory, I stayed in Old Town away from the nightly sirens of downtown that reminded me of Baltimore. Every time I hear a siren, even today, I think of those days in Baltimore when sirens were part of the nighttime environment.

My teams were at the McDonnell Douglas Delta plant, TRW (now Northrup Grumman Space Technology) in El

Segundo, Aerojet (now Aerojet Rocketdyne) west of LA and in San Diego at the campus of USC, SD.

LA also had the Aerospace Corporation (think tank spinoff of TRW), a place I frequented seeking the latest Space Engineering practices and techniques. I became friends with an east coast born manager named Bill Tosney, one of the best Systems Engineering Managers in the business for my money. His recent Aerospace books are standards of the industry and I was proud to be one of the first to receive courtesy copies of them for my use at NASA. We had the same desires to push organizations to adopt new engineering tools and practices that resulted in safer missions and applying past lessons learned measures into the entire future spacecraft building process. A similar business company in Palo Alto fascinated me called Failure Analysis Corporation. They too studied engineering failures and even did and video analysis of the President Kennedy Dallas assassination with computer graphics that drew wide public acclaim at the time.

Hughes Aircraft, McDonald Douglas, TRW, Aerojet, Northrup, Rockwell and of course Lockheed were the big players in LA. And up the road, VAFB was home for NASA Polar orbiting spacecraft. VAFB was originally planned to be another KSC Shuttle launching station, however, funding was never given for this to take place. NASA during my years was always fighting to maintain its low budget of less than one per cent of the whole Fed Gov pie. In fact, they needed to compete with the VA for funding, since they were on the same line item budget. Why, I never found out, but surely a big oversight. But at GSFC with, who we called St. Barbara from Baltimore, we had a fearless fighter in our corner. GSFC is surely going to miss her when she retires next year. God bless Senator Barbara Mikulski; another person whose statue needs to be placed on the BW Parkway to DC as thanks from all of us who worked at Goddard.

My guy was resident at the Delta plant where they had the missile building business down pat! Every detail was covered and checked again and again. Of all the high-risk areas for problems to occur, I always felt confident that they were prevented from happening here. Even today, the Delta Team has a high reliability record and new launch pads were built at KSC just to accommodate their launches.

TRW in El Segundo in addition to my Assurance Management team had a NASA GSFC team of engineers on the TDRSS Project, there for almost 10 years, I always wondered if they were there TDY and getting additional pay to live in LA where it is a high cost of living city. We needed to pay our people upwards of 50% more than east coast personnel and they still complained that it wasn't enough! I had the highest respect for TRW since reviewing their launch site manuals and documentation at KSC. They were the finest documentation instructions I have ever seen, and all launch teams followed them to the last detail.

My original guys in LA were there before I was even hired at NASA. I was always surprised by the NASA job longevity that most of the folks there had. Some were going on 30 to 40 years with the same job, unheard of in the commercial world, even with the DOD contracts. Turnover in the high technology business was like lightning…I worked for almost 10 companies before going to NASA…quite common for 30 years in the business.

The west coast especially the LA area was undergoing many federal investigations of large-scale Aerospace Contractors at the time early 1990s. The business had certainly changed. Unheard of practices became a concern in high technology. One that bothered me as an assurance professional was the falsifying of product inspection and testing documentation. This was a very serious offense one that involved customer safety and survival. I, as a seasoned manager, always used "a trust but verify approach" like President Reagan did with the Russians during his historical tenure. I made sure every member of my team knew where I stood on these matters and part of their job was to be on the lookout for any unethical practices that may be going on in each one of their oversight factories.

One Aerospace contractor not covered by my teams had over 15 personnel indicted for taking part in falsifying test documentation and some were in the Quality Assurance business. After looking at my new West coast team's resumes and applications upon returning to Goddard, I found out the two of my guys formerly worked at the same employer that had federal indictments. Fortunately for all concerned, my guys were the good guys and not involved in any way with the shenanigans uncovered elsewhere. My Team all consisted of good guys with one exception back at Goddard.

Process Quality Checklists - one of the quality controls that we developed for field use was the use of site checklists. These, too do, daily, weekly and monthly operational checks were to be performed by on-site personnel to assure that process measures were in place and operational. When I did my round robin site surveillance, I would check the logs and act like us as a second look to make sure that the processes were "in control"! At KSC, I was impressed by the checklists created by TRW and adopted there is instead of reinventing the wheel. Process controls were a NASA requirement to assure safety and quality compliance.

All off-site personnel were given annual quality reviews and training. I myself did training sessions, one on ones, with my field team members. I think the sessions brought us all together with a common mindset and it was time well spent and appreciated by all personnel. After Columbia's fatal flight, all processes were reviewed, and additional preventative measures were put in place from the report lessons learned recommendations. The entire NASA Spacecraft Program was reviewed from top to bottom with all the recommendations resulting from the final report on the cause of the disaster. Everyone involved was re-trained and given all tools and techniques to provide additional layers of flight assurance to an ever-evolving critical design, build, test and launch process.

I last managerial trip to KSC was in 1998. The glories of the 1960s were now slowly becoming memories of a generation gone by and mostly... retired.

The one thing that bothered me about Martin Marietta was the reputation rumor that the average retiree only received 12 to 16 checks after retirement. Martin just burns them out as the old-timers said. I tended to believe it after witnessing almost 5 years there at a very high stress level operation at least in the Baltimore facility. I fortunately only worked for there for two years with another almost three in Orlando.

Today's me generation perhaps even my own offspring did not get the nuclear hardening to survive such Theory X management style as most of the workplaces were in the years that I worked. Fear was the most powerful motivator for my culture and the easiest one to use as a manager. As an MBA, I studied and even taught leadership styles. And until the digital age evolved, most companies in the United States operated on purely Theory X philosophies. For introducing Theory Y was almost like a worker's Emancipation Proclamation being passed by the Fed Gov DC powers.

I perhaps told you the story of my own daughter's workplace situation. As a computer guru, she complained to me that she had not received a raise in a few years. She worked for a Virginia capital law firm of hundreds of attorneys; whose main business was to keep the tobacco management out of jail. I recall most of their blatant lies over my lifetime, especially the famous TV Kent cigarette commercial of the 50s, when they made smoking entertainment, like eating cheesecake by the seashore on a lovely summer day! Well, for most of those now departed, KENT smokers, mostly women those days, little did the girls know that they were sucking on an asbestos filter every time they took a drag! So much for that micronized filter, now back to my daughter's problem! I told her that on the next workday, to just go into her boss's office, close the door and say I have a message from my father, "that the Emancipation Proclamation was passed and signed by the President Lincoln in 1863. Slavery was abolished!" I personally have not had a raise in years! End of his message! What would you do in such circumstances! If you're best computer person did such an act? Yes, you like he did, would tell her it would be fixed immediately after that incident. She was given a salary adjustment regularly nearly every six months. She also, traveled first class to their new European offices, training new lawyers and computer people. And she was treated royally right, up until she retired at 32 years old to become a full-time mother. Oh, I reminded her after she told me her retirement news, that her mother beat her, she retired at 20!

Speaking of retirement, when I finally retired from active service at Goddard in 2008, the DC area had 78,000 openings. In the bedroom community of Laurel, MD, over 35,000 new single-family homes were being built for future Fed Gov workers. I can honestly say that in my 25 years in the Baltimore/Washington arena, there never was a recession. New towns were even built, one called Columbia, a commute from DC workplaces, one that was "environmentally friendly"! The little villages that made up the town with almost camouflaged settings to the new resident or visitor. I had a very difficult time even finding a gas station among the many mazelike shopping plazas. I missed my Baltimore County lifestyle and after one year returned to more familiar surroundings.

17. The 1990s: New Aerospace Business Fears & Intervention Measures!

My top contractor boss asked to see me one day and handed me a resume. Look at this he said perhaps you can use him…an Annapolis graduate who is looking to come on board at Goddard. Yes, it was an impressive looking resume, so I called him in for an interview. The DC beltway inside and out is a honey pot of opportunities for what I call tramp executives. These are what I also call the operators who look for job opportunities to feather their nest. The higher the resume qualifications the more one needs to screen their backgrounds for verification. One of the most difficult problems at the time was industrial espionage, a practice of infiltrating organizations finding their strengths and weaknesses and uncovering anything of value, especially technology inventions or company practices and then selling them to the highest bidder! Industrial spying in the aerospace business is also a subject not to be ignored, especially by management.

Guess what I found out later after just a few months of this referral by my boss came along. You guessed it! One of the very best Beltway operators in that business right here in River City! I had my boss dismiss him, after finding some incriminating documentation in his possession, since he was the one who initially caused the problem. Another lesson learned for me in a new, for me, Baltimore/Washington arena.

In my later years another problem that NASA and the DOD needed to resolve was counterfeit parts. As a frequent flyer, I was very interested in how to deal with this new threat. Can you imagine flying in an airplane that used counterfeit parts for its maintenance? Bad enough in our automobiles and household appliances, but for Aerospace use a definite no! NASA had many seminars all that I attended on the magnitude of the problem and what we as managers could do to interdict their use. Such preventative actions such as procuring parts directly from the manufacturers and conducting 100% inspection and pre-testing was immediately put into place. However, these counterfeit parts were very hard to detect. Even to their identification makings and external looks, it was hard to determine the good from the bad. X-ray and other tests like loose particle detection (Remember the FOD story I told you about at Martin as one of the biggest fears in a test pilot's mind)! The same fear our astronauts had as John Glenn often said his biggest fear was that the job went to the lowest bidder! The counterfeit parts business is in the billions and is a worldwide problem, still not solved to a safe level.

The cultural changes that now appeared in the workplace was another concern of mine. As I found out NASA employed many foreign nationals in all its Centers. Engineers with green cards worked all over in DC. This started back in the 1960s when then President Kennedy allowed them to work in the NASA Agency. And I am not talking about a few, I met many in Manager's positions and some of my later contractor bosses owed allegiance to foreign countries from their middle east to Asian homelands. But, recall my early experience in Huntsville when NASA started with over 165 German Scientist back in 1958…some of whom were high level Nazi military officers. These people made up much of the personnel who at the German launch site at Peenemunde launched V1 and V2 rockets at our allies in Britain. One of the first books I bought and read in Huntsville in 1960 was a best seller written by German Army General Durenberger, Commander of the German Cape Canaveral called Peenemunde. These same folks now formed the MSFC and helped us in our quest to be the first to land on the moon. Most did become loyal US Citizens over the years and loved and respected our system of freedoms and government of the people, by the people and for the people.

In Flight assurance at NASA we were always dealing with problems. And in an engineering manager's job, was putting out fires as we referred to them. One of our chief concerns was determining the root causes of mission failures. And, never to be taken lightly, was the factor of sabotage occurring in workplace by strangers or in plants by someone who may have a grudge against a company, its practices or its people. NASA Assurance Management's job was to make sure things were done right and that every operation was checked and signed off before the next operation began. One can never be naive in believing things are ok, one must look for things that are suspect and not ok! The high technology space business is one with high visibility on things done wrong by the media especially such an adventurous agency as NASA. One of the best preventions done at all our sites was the introduction of video cameras on 24/7/365 showing the spacecraft being built and watching it with live coverage around the clock. Even KSC Space Shuttles were modified after Columbia, to place on several board cameras that showed us 360-degree liftoff photos like never before.

I became well versed in accident prevention and Assurance Management even before going to NASA. However, I was able to provide prevention measures knowledge to existing and new projects that became standard practices. Such measures and proven practices as our home brewed versions of Mission Success, Barrier Analysis, Benchmarking, Dr. Deming's Rules, Crosby's Quality is Free! Even Six Sigma and Electronic Dashboards and Scorecards for illustrating performance and weekly management reviews and reporting were frequently used tools we applied to everything we worked on. Later, even with NASA Spacecraft, new software design reviews, program plans and even engineering analysis covered cyber vision prevention practices that when first addressed, were laughed at! As they once they laughed at Hershey too and Apple and a new company called Microsoft, when it first launched in Boston in the early 1970s. Bill Gates co-founder today is the richest man in the world with over $76 Billion in the bank! And Steve Jobs was a billionaire before going back to rescue Apple and give us some of the greatest new technologies ever conceived by mankind…I just love my Apple products and have so many HP computers I lost count…as soon as I fill up the hard drive I buy another one, my latest HP Pavilion

Touch Smart Notebook touch-screen laptop (with wireless remote keyboard and mouse) is what I am using to write this book! And after buying my Apple MacBook Air Laptop completed it with greater speed and fewer errors. Jobs was way ahead of the PC pack in my book!

Photo 17.0: NASA's Johnson Space Flight Center (JSC) Houston, TX Large Front Entrance Sign…Designed by the President himself…LBJ!

Houston, TEXAS USA! - Home of **"THE LYNDON BAINES JOHNSON SPACEFLIGHT CENTER"** (JSC) and NASA Mission Control and the NASA Astronaut Training Center. I capitalized the center sign in the front of the entrance at JSC. Then President Johnson wanted the letters to be as big as the ones in St. Peters Cathedral in Rome and they were…I verified that myself. On my moving to the Boston area in 1967 my direct boss Ed Rickley, after just a few weeks of my coming on board, joined a new organization in Cambridge, MA outside Boston, called NASA. Ed was a very sharp engineer and he thought that as was President's Kennedy's intentions that the Mission Control Center would be near MIT/Harvard in Massachusetts. However, once LBJ became President,

Texas would be the new home for it, specifically in Houston! I guess Ed would go on to be a Federal Transportation Authority employee since Boston, without JFK, lost the new NASA operation. I have a book on NASA JSC Houston that shows inside the front cover, the land before NASA...an empty field of grassland on a college campus and inside the back cover...after with all the new buildings on it. The vast complex now contains a sight to behold and illustrates the explosive growth of that era of reaching for the moon and the stars. Fortunes were made for all concerned, the cities, the states, the companies and the employees...just amazing growth and an entire new industry was created and born!

NASA: Johnson Space Center...My contractor's story: Play Just to Play!

I was called into my boss's office one day when he requested me to join a Red Team at JSC on an RFP (request for proposal) effort company initiative in Houston, TX. I had winning skills in writing proposals for years and they wanted my expertise in this new business write-up. My employer had work in both GSFC and JSC and they were seeking to broaden their base at JSC. A new job was going to be bid, and I was selected to go TDY to help win the job or enhance the proposal and make sure all the NASA contract requirements were addressed.

I was amazed to find out that several companies were housed in the large office building in the same off-site location although cypher locks kept the co-mingling from occurring, I was doubtful that this could be an effective security measure. I after a few days of dabbling in corporate BS, I asked a key question of my corporate hosts having been involved in many proposal efforts in my career. "What are the chances of winning this job?" Well they replied, "Around 30 to 40%!" At that point, I said you don't need me for this exercise, I will send you down one of my writers and made my plane reservations to return to my job at GSFC. These exercises were a waste of time in my opinion, but they were the types of games corporations played just to be in the arena! In the end at JSC the new work was awarded to a competitor and my time was better spent on important spacecraft project matters awaiting my return.

This game was also played later ARC (Ames Research Center) in Mountain View, CA my old stomping grounds near HP. I was sent to gather some G2 as we called it a reference to the military term organization for intelligence. I welcomed the east coast winter break for some R & R (rest & recuperation) in the great USA, city by the bay, San Francisco! Ames was a special government facility going back to the early days of WWII. As a special laboratory resource, Ames had a one of a kind, state of the art, wind tunnel testing facility that was used by all aerospace and airplane producers. I rated the operation as a Class A facility as the personnel and organization was top notch! This new contract was eventually awarded to a local ARC contractor as it was very difficult to unseat an incumbent or for one NASA Center contractor to beat Center's served market supplier. Only in the small business set aside and 8A business sectors could inroads be made. Each of the 15 NASA locations had their favorite son contractors, some in place for many, many years.

NASA Contract renewal that I was on occurred every five years. I participated in two wins in a row that took my employer out to almost 40 years of incumbency. As you have noted, the Beltway Bandits as the DC area contractors are called live and breathe federal contracts. Some are called "set asides" for these are the most coveted types. Only experienced players participate in these...often called the insiders due to special skills, experience or track records. Others are the low bidders, now called "best value", the kind John Glenn feared on his first space mission! Lowest price today, seldom is a guarantee for victory as we shall see later.

When I first came on board, the 5-year contract I worked on was up for renewal...looking back the reason they were looking for outside expertise for the heat was on for experienced problem solvers with industrial experience. For many spacecraft missions were backlogged due to the recent Challenger accident! Through lots of hard work and high customer ratings (like a report card every month, quarter, half year and yearly) my employer retained this contract for over 30 years, since NASA's beginning. Competition was getting more difficult as aerospace contractors found this work to be highly profitable many times achieving additional fees for work well done. I had worked on proposals for many years and achieved a significant level of winning for my employers going back to the 1960s. This one would be no different, much work effort would be given to provide the best technical response from all the other bidders. After 30 years, however, especially in the DC area, hostile forces are always part of the game! Jealously and strong competition are forces always undermining the best of intentions. We were no exception as you shall soon see.

For this proposal effort, even I was impressed by the technical response. Each Department being supported from Finance to Engineering Laboratories, Project Engineering Teams and Specialty Services were addressed and meticulously covered. Sweeteners were weaved into each effort where better value was emphasized. Hours of effort in writing, reviewing and preparing for even "oral" presentations resulted! I was particularly impressed by my Program Manager's boss over one hour "oral" presentation! For over 30 years, he had his act down to an art form. I said to myself few managers could do the job he did at this task!

No competitor did a better job! We won 5 more years of work despite the many enemies I found out later who wanted us to lose. Five years later it was quite a different story! NASA new management forces in charge wanted a new player since many of the first generation were retired or planning to retire; the opposing forces were gathering like sharks feeding time! My employer knew a fight was for companies from other NASA Centers were opening temporary offices close to our base of operations and even holding open houses for potential employees if our team lost! The contract had grown to that magic figure of $100,000,000 and for most of the big-league players enough to "turn the crank" to try to win it!

My job this time was not just a writer, but a major contributor! And this work was to be done in addition to my normal duties that were equally demanding! I also, by this time, had many jealous colleagues who wanted my team and my management slot since we were the leading scorer in all customer reviews. When in the beginning our proposal, team seemed doomed by new unskilled management and the company top guns admitted that the wrong people were in charge and did not know what they were doing.... Headquarters decided to bring on the experts!

One Saturday morning as I approached my Reserved Parking slot, I noticed many new BMWs parked nearby. The "A" Team arrived with their new cell phones before they were in vogue and computer laptops loaded with sophisticated software ready to do battle. A joint team meeting was called with all proposal writers and the new players were introduced. These new folks were expert spin doctors in doing their jobs. A consultant was now calling the shots and he was a big guy to say the least. Someone said his brother played with the NFL Pittsburg Steelers and he looked like he could too! We were all told to take orders from him as the old management were sitting on the sideline and kept their mouths shut! I was immediately relieved and soon found myself face to face with this new Joint Team Leader! He told me that he was from Pennsylvania and yes, his brother did play with the Steelers. I replied that I too was originally a coal cracker from PA! We of course immediately bonded and became close friends throughout the several months ordeal.

Let's sit down and tell me what you do, he said, as he was doodling on his large writing pad. We support the Space Shuttle program I replied, and he started to draw the shuttle. As I described my team's roles he added to the graphic more and more until it resulted in a mural. Is this what you do he asked? I looked and said Wow! Yes, exactly! Ok, he replied let's start writing! Page after page covered the intimate details of his mural. He was revealing a success story one describes as mission success, a powerful proposal differentiator to be seen paying off big time later in the coveted award-winning process. I worked many weekends and overtime hours on this proposal. I tallied over 750 UCOT hours that were unpaid. Oh Yes, UCOT, what a Beltway scam for those of us that never got paid!

UCOT, I wish I could find the Beltway SOB who invented it! We all hated it! And we worked free for our contractor employers who used it to help win and keep their highly lucrative contracts for years and years! It worked like this, if you worked over 40 hours a portion would be "unpaid" up to a certain amount! Even this company top management used it for doing proposals! I found out later that the higher ups were not operating under the same rules and received full pay even big bonuses if winning the job! As a Director during my career the awards often exceeded 5 figures and was tiered down from there for all contributors.

Upon Award decision day, we got the word.... **we lost!!!** The contract was awarded to a KSC Contractor for a lower price! All my hours of work went down the tubes…was all I was thinking about! When my boss called me and told me the news I replied, you mean we are not going to protest the loss? No response followed! I found out later that a VP from Corporate was "bullshit" on hearing the bad news! This three times, shot down helicopter pilot in Viet Nam started the appeal process! Now everybody inside and outside the Beltway knows there are impossible odds in overturning a Federal Government contract loss! However, as I often repeatedly learned that the impossible is easy, the improbable is hard!

After we obtained the best DC Beltway law firm for a significant up-front cost, the protest process began. The hearing was before a newly appointed female judge. Almost like my last divorce hearing that I finally won with my lawyer who I called Mister Ugly (differentiated from the good, bad ones) but never lost! Each page from our technical proposal was scored and graded compared to the winner contractor. We got a 10 out if 10 score on every page. The winner a 7 or 8, but no 10s! Case closed she said, the

incumbent contractor wins!!!!! So much for my first 10 years at NASA!

Later, however, the Federal Government got one up! To the best of my knowledge, we were the first and only Contractor to overturn an awarded contract at GSFC. Next contract years later, the NASA contract evaluators were given extra training and our contract was broken down into little segments and called an 8A set aside for small business players only. No more using just big-league players at least not at the top of the heap!

My own employer had approached almost 35 years with the same contract somewhat a record in my book. And for some of the employees, it was their only employment for that long period of time. This was unheard of in my commercial sector experience over my 50 some years as contracts went back and forth between leading industry producers. And even more so, in the aerospace business, as there are not that many competitors to contend with.

The later contracts were often supported by a big-league player in what was called a mentoring role or minority partner, who helped do the preparation of proposals for the 8A leading contractor. The long-time big-league players were soon becoming history as industry consolidation was the new game almost being conducted like a packman…soon only the strongest players would survive.

Cape Canaveral, Florida, the final destination of my round robin trips ended at KSC! My team personnel were located actually on the Air Force side of the large sprawling complex. One could not enter the base or KSC unless a special badge was issued by Security. You had to be also on the KSC Security list and have business there or no entry was made. I endorsed these rules for earlier when I worked on Pershing in Orlando and travelled to KSC, I thought the security was rather lax…especially after spending my Army time at Redstone were you were greeted by MPs at the entrance gates…these places need to have the highest security possible. I found out after the Challenger Disaster, a new US Air Force Commander made his own assessment and added many additional barriers to the highly secure complex. Many protest groups tried to enter to no avail since these added precautions were instituted. One story, I recall being told was one protest group tried to cause problems with a launch and were told to gather at a certain part where they could do their thing...it was near an alligator infested area of the base and they got out of there soon, once the gators started protesting too!!!

I went to the old Saturn launch pad one time and was surprised by the signage in place indicating "Government Property – Salvage in Place"!   I thought my, oh my...a place where destiny occurred discarded as obsolete...I could see pioneers like Von Braun and the Mercury 7 Right Stuff Guys and Gemini crews and finally all the Apollo missions, especially 13; all that happened here now gone with the wind too! A certain gloom was felt when I pondered this hallowed ground now bring cast off like trash...why wasn't this preserved like so many of our USA treasures? A historical site period that should never be forgotten!

Just a few words on my thoughts regarding what was called by many beltway folks as the social engineering experiment. Since I worked almost 25 years both inside and outside the DC corridor, I think I can speak with some authority. I recall on my first visit to the DOD in DC in the late 1960s. And watching, while the US Mail was being sorted and distributed to a long series of adjacent managerial offices. Individual Secretaries attended to each office. They were at desks right outside each high-level officer or civil servant's office doors. I counted at least 15 to 20 offices that were receiving the daily mail.

The mail arrived at each Secretaries desk, who then sorted through seeking her boss's correspondence and placing it in his IN basket. Afterward, the mail was passed to the next Secretary, followed by the next and then the next until it was finally emptied. The Senior-level Officer I was visiting saw me witnessing this daily ritual. When I got into his office, he closed the door and smiled at me...saying, "Can you imagine if all those people were on the street?" Such was my first exposure to the bloated bureaucracy they call DC! Everyone had a kingdom and held some level of power that started with military or civil service rank and ended at the top with 535 publicly elected and in most cases career politicians. Coming from private industry, it was a complex system that could actually "blow one's mind" in trying to figure out how it worked to prioritize and eventually pass any of our important business or new laws. I will try to show you one example of the system at work in contractor managerial selection.

Affirmative-action victim - on all federal contracts there is what is called the evaluation period. In our case, at the time, it was conducted every six months with the changing of the NASA govies old guard occurring due to retirements and edict to promote minorities; the old space culture was changing rapidly. This occurred throughout the 1990s, and is still going on, only at higher rates. Some of my best friends at NASA were minorities, and like myself of Irish/Italian heritage! Even my minority NASA boss was an ex-Army officer, who vision for a better workplace was the same as mine. Fair play was always on my mind whether any race, color, or creed was concerned. During a new contract evaluation, the new NASA Director, an ex-contractor like me, who was a minority, scolded my employer formally in annual Contract Award letter, for not having enough minorities in our management ranks. We were penalized monetarily for this situation with the proviso that if corrective measures were taken, lost award funding would be restored. And since a new minority manager replaced my Contractor boss too, who had the job for over 30 years, swift action occurred. I was almost in the management slot for 10 years and had a team whose performance record was second to none! Our team ratings were even a perfect 10, during one evaluation period. And we consistently scored over all other departments, however, like everything in life, it did not last forever. And 10 years seem to be almost a record! To give up my management position, I was offered by NASA, a position until retirement. And, fortunately for me, a

213

corporate buyout occurred by my contractor firm, since they were not eligible to bid on another contract. The government went one up, after getting even, so I was lucky to take the offer! Two people replaced me under a new small business contractor. Both were minorities, one male and one female. My organization structure, however, was still in place even at an enhanced level. Years later, the structure was in place after all these years. However, the players did not have the same seasoning as our former Task Leader team did. The first big negative event occurred after I left was with the NASA GOES N Satellite being built in Silicon Valley in California. It seemed a new player replaced my ex-Marine Vietnam Vet who played the role as a then tunnel specialist! He had the job for several years without problems. I do recall his departure, and later a Monday morning, I received a phone call saying he would not be going to work today or for that matter any day in the future, as he won the $12 million California lottery. It seemed in his absence; the new player was off duty during one of the evening events, when some satellite retaining bolts for keeping the large structure in place were borrowed for another job. This unauthorized action resulted in the over $25 million spacecraft tilting over and collapsing on the shop floor! See the NASA Report on GOES N for further details. I too was in a new role called Mission Operations at another small business contractor. All of us contractors suffered from this tragic event as no profit or gain sharing checks were given out that year as our ratings soon dropped several levels down.

Search for photo....spaceflightnow.com...good photos!!!

Fig. 17.1. Three NASA Amigos with Gene Krantz, Apollo 13 fame, movie & book , "Failure is not an Option" mantra. L to R: NASA Joe Osche, Gene, Jim Suraci & Me.

PS Shortly after this photo Gene (a fourth degree K of C member) contributed all his NASA Memorabilia to his local church and they were lucky he did! For years later, he lost all his possessions and home in the disastrous hurricane flood in Houston, TX.

I'm pleased to say that Gene is alive and well in his adopted Texas home and still on the speaker's circuit. And according to our recent correspondence, he is planning a new book for our reading pleasure.

18. Celebration Time: 35th Anniversary Party of Moon Landing: July 20, 1969 – July 20, 2004

One unforgettable time in DC was the big celebration party that my fellow colleagues and I almost missed. It was held 10 miles away in DC at the Aerospace & Space Museum in downtown, on the famous Mall. That afternoon, I called several colleagues and asked if they were going to the big Party. Most hemmed and hawed trying to duck the occasion as we had many times been invited to special occasions and never attended until this one. Hurrying to make the DC bound train from Greenbelt it got us there just in time. After a double screening by Security, tightest I ever saw before because all living Moon landing era astronauts were in attendance, including the keynote speakers Walter Cronkite and Neil Armstrong. The President sent his best wishes and did not attend because it would take away from the real heroes...NASA's Astronauts!

We forgot to eat and were almost famished when suddenly a catered dinner was served, and it was the best of h'orderves, top shelf done the White House Caterers...simply the best DC could offer. At the same time, mobile bars appeared with an array of beverages to compete with the best DC watering holes, even in Georgetown! NASA people were mingling in and reviewing American history with past heroes, still alive and discussing their Space experiences. After devouring all the fit for a king food, came the carts of desserts....and after diner drinks, all top shelf brands. My friends and I were overwhelmed by such a display of opulence, but in attendance were the best space performers the world has ever seen, together on a very special night. We adjourned to the auditorium for the main event, a movie and talks by the Apollo moon landers. It was an occasion never to be forgotten. My chance to finally meet and shake his hands with the greatest USA Space heroes. Lines were forming everywhere, and I asked my friend to take my photo with Walter Cronkite, the honored guest. Later over the weekend, I showed my Dad the photo and he could not believe I met him. I told Walter, it was a pleasure and honor to meet one of my favorite personalities of the great space program. He nodded and shook my hand firmly. I asked later how much did this party cost and the figure was close to $300,000 including the great souvenirs given to each attendee. I treasure mine and occasionally look at them and remember one hell of an occasion in our Nation's capital, with all my heroes, in Washington, DC!

On Saturday, September 6, 2003 at Lockheed Martin Space Systems in Sunnyvale, California, NASA GSFC's NOAA N Prime satellite a serious, costly mishap incident occurred. While assurance personnel were off duty, the entire satellite fell over to the shop floor, off a turnover cart. I was no longer associated with the Assurance Management Office field organization at the time, having moved on a few years earlier, to the new Mission Operations function. The new team of management was now responsible; however, I was working for the same contractor, and all team members suffered. The reason was that the new contractor for oversight lost its annual award fee and as an employee, I too suffered from the loss of this incentive award fee. As we said at the time, perhaps the cheaper, better, faster philosophy just did not apply here. From 1989 to 2003, this never occurred under my watch and even later with my replacements. Switching from large-scale big business, space firms, who for almost 40 years held these mission-critical responsibilities, NASA procurement now focused on a cheaper, better, faster, mantra with small minority owned businesses filling the voids some with little or no track records. In my humble opinion, these new trade-offs did not, over the long run, save NASA money. For mishaps like this one, failure costs ran in the billions when counting the Columbia and Challenger shuttle disasters. A NASA Mishap Final Report, dated September 13, 2004, is available online at NASA.gov. The report indicates that oversight was negligent and in-plant quality controls were lacking as the root cause in the mishap.

Several reports, done quite professionally as are all NASA's are available online. I have listed a few for the reader's future follow-up. They are:

o Columbia, CAIB Report Volume 1, dated August 2003

o Challenger Report January 28, 1986, Presidential Commission on Space Shuttle Challenger Accident of June 6, 1986.

o Apollo Report, April 5, 1967 on KSC Fire of January 27, 1967.

o James Webb Space Telescope Comprehensive Review Panel Final Report, November 10, 2010.

Cost of Quality (COQ) is often referred to as a measure for the producer of a product to determine their real cost to make one's product offerings. As we referred to earlier, the original digital watch business that used LEDs & LCDs at first, was highly profitable! However, as the volume went up, and high field failures too, the manufacturers soon went out of business! Why, because for every two watches made, one was for warranty replacement. So, eventually costs skyrocketed and sales revenue fell thereby causing high profits early in the game run to become now, big losses due to poor watch reliability. As I indicated earlier, after I digested all the Columbia Accident Report and the Presidential Report on Challenger, NASA's Cost of Quality reached the $Billions. So much for the cheaper, better, faster, ideology proposed by the newly appointed Administrator, at the time. I was amazed with all my fellow managers who supported this initiative and that once the new administrator left office his infamous words immediately disappeared from the limelight.

After Columbia, the new mantra emphasized both Quality and Safety practices. And although not formerly a college or technical school course, the same managers quickly adopted the words. Some were even to become Quality Champions a subject they knew little or nothing about. World Quality

Standard ISO 9000 was the new flavor of the month initiative. Interpreted, by those same now bandwagon managers as being just an effort of documenting what you do and how you do it (even if it is wrong, I commented?) will suffice. Some even said that it was just common sense, nothing more, showing their ignorance for a practice they just never really understood. This was their simple explanation of what really was a United States Military Standard entitled Mil-Q-9858A. A mil spec that I have been using from my early days in the DOD arena. What really occurred was that the European market copied it and retitled it with their own format. It was a way to stop USA made products unless you had European certification. And then they sold it back to American business as a new process to produce products and achieve higher levels of quality. Smart companies in the DOD business for years then, did not need to change anything and already complied with the new so-called standard. Between new quality gurus like Dr. Deming and Phil Crosby and the new European ISO 9000 groups, the USA finally adopted quality measures that have been around for years and invented in the USA.

In fact, several aerospace companies in California got together and decided that the new standard was not tough enough and immediately adopted over 30 new controls resulting in a new Aerospace 9000 called AS9000. As I said before, I saw the beginning since it was developed by the West Coast committee for months. I tried to get NASA to adopt it immediately only to be pushed back by forces in the DC NASA Headquarters. "We are not adopting AS 9000, and don't bring it up again or else I was threatened. Years later, I read a HQ Memo that said NASA was adopting AS9000. Adopting like a child, I replied, since they had nothing to do with bringing about its birth! AS9000 became the NASA standard after it was forced by contractors to do what all the others had adopted for many years. And of course, the Columbia disaster sure did help pave the way tighter quality controls.

One area that, in my humble opinion, NASA shines is the high quality of what they called Mishap Reports from the Presidential Committee Report. They reveal striking photos the Challenger accident, to the latest shuttle report on Columbia, all are top-notch write-ups! This effort, however, added to the high cost of failure that we discussed earlier in that the text tens of thousands of man-hours to produce such documents. And the teams of experts to put them together, in addition to, implementing corrective and preventative action throughout the process. All act as a multi-layer of total expenses often overlooked as a total cost factor. Even the personnel retraining, new process control modifications, re-inspection and re-testing efforts and delays in the mission, all at up to significant added costs.

Works Still in Progress: The James Webb Space Telescope (JWST) The Next Hubble!

Photo 18.0: James Webb Space Telescope Mockup in front of GSFC Headquarters Building with the NASA Design Team members, including yours truly!

My last NASA Project challenge was on the yet to be launched JWST scheduled for 2018 launch (now 2021). The new deep space telescope was to go beyond the current HST mission and go almost one million miles from planet Earth to what was called L2, far from our planet! The HST mission launched in 1990 is still ticking after several daring missions to maintain and service it. As a Sr. Systems Engineer, I was assigned with a NASA Senior level seasoned team to help develop design requirements criteria for all the hardware and software used on the latest spacecraft. Part of the JWST mission one million miles from us, where it is completely dark, is to explore space from what is called deep space.

My job specifically, was to join the NASA spearhead group called Systems Engineering to develop the satellites engineering documentation that would be used to design, build, test, launch and operate the complex state of the art, new bird, as we called them. I must admit that the team I became a part of were the best engineers and scientist that GSFC could muster. All had years of NASA satellite experience. And, like myself, were my post WWII generation doing their "last hurrah" work they would never experience again in their lifetimes.

As in any team assimilation process, I need to earn my respect among my peers. And being baptized in the GSFC Flight Assurance Management Directorate side of the business, making sure things were right being of prime consideration. The NASA boss was a guy that we, weeks earlier convinced with our canned "Mission Success" PowerPoint presentation that he needed experienced engineers as part of his new project! I was first assigned the James Webb Space Telescope (JWST) Systems Engineering Plan to review and embellish it with mission success elements and make it reflect all the lessons learned including both those from the Challenger and Columbia disasters.

After initially reviewing the Plan documentation, I recommended extensive re-write and updating to these new requirements. It was big job and needed to be done yesterday! I had some great help doing this job with a small team of engineers that helped me on the AQUA mission ground systems requirements documentation validation exercise. We, in short order, filled in all the holes and beefed up the Plan to the latest standards and mission success elements!

After completing this several weeks team effort task, the next assignment I was given was to review the JWST Software Development Plan. Lucky for me and NASA software was a technology that I was currently up to speed on; hiring many of the new software engineers at GSFC for the Assurance Management Office function. By this time, me and my colleagues made our mark, the systems engineering team adopted us and we became a viable part of the new JWST Mission development team.

Now let's digress and discuss how I saw hundreds of millions of dollars made inside the DC Capital beltway over the many years I worked there! Starting in the 1980s, small government contractor business, in a large part sponsored by the leading US Aerospace industry giants mentoring efforts, circled all the many NASA Centers. From the east coast GSFC, to the Cape, KSC, to Houston, JSC, Huntsville, MSFC, Michaud, LA and Cleveland, GRC, to the west coast Ames Research Center, White Sands and Vandenberg even in NYC all NASA Activity locations.

These new and in many cases minority owned enterprises called 8A companies sprung up and took all the small contracts (under $100 million) and did them cheaper with their low overhead expenses than their former players. For the big players like the Boeings and Lockheed, Raytheon and Martins seldom even "turned the crank" for under $100,000,000! However, as I said earlier, they were required to mentor or sponsor these new smaller companies as part of larger contract orders. Who ever said that DC could not develop new businesses…for thousands of new ones appeared or as in today's popular chant to "redistribute the wealth!" Some of these new companies were partially owned as far as the law would allow. I followed and even worked for a few myself! To make a long story short, I tracked four of them for years all sold before I retired for around $200,000,000 a piece! Yes, $200M! That's over $800,000,000 with just these four! CEOs of these four became multi- millionaires overnight and while they were still young enough to enjoy it! All were minority-owned businesses. Some of my friends who helped them, even as Director's, were given pink slips and zero dollars so much for sharing the wealth! And so much for us helping these new companies grow and maintain compliance with the new government contractual requirements

One CEO, of Indian extraction, I personally knew very well; he always visited me on his salesmen rounds gathering G2 for his ever-expanding business. As some say, they did work hard for their money, but I also say didn't we all. As a manager and worker bee, I helped my employers make millions from my individual work efforts only to be given in the best of times several thousands of dollars, in addition to my regular pay. This did not even cover the OT pay I worked! I never received compensation pay for so-call UCOT (Beltway acronym for "Uncompensated Overtime" a term probably invented by a hot shot bean counter at the time)! Back to this CEO, he was a very smooth operator and if you told me that he would someday before I retired would walk away with $200M, I would never believe it! But he did after just over 10 years!

All over the beltway, these new insiders were like me reaching retirement age and buyouts and mergers were occurring everywhere! The new Beltway Bandits JV Teams were now becoming the first string! NASA was now going on 50 years of age and if you joined in your 20s you were now hitting your 70s! I recall attending the 35th Anniversary Party of the moon landing in DC when attendees were asked to stand up if you worked on Apollo, my good friend Joe Osche and two others stood tall in a crowd of hundreds of NASA attendees. I myself was 65 and my WWII Dad was still alive, in his 80s, when I told him I wanted to retire! "You are too young", he quickly replied! So, I continued to work until he passed away at almost 90! Joe who has his own company, wishes that he was granted the same opportunity that all these new CEOs had; but the rules were not made for him to become a player in this new beltway game! No extra points were given to this European offspring guy from Pittsburg, PA! As we said many times America is the greatest country in the world! Where else can regular folks be given opportunities as they do in this good old USA!

My NASA Heritage was discovered on a visit to my Mother's cousin Dr. Charlie Lacurdo who retired from GE Valley Forge, PA. Charlie was not impressed by my credentials or NASA war stories, for he was a pioneer in the space program from the beginning. Well into his 90s today and still with an inquiring mind, he wrote the reentry equations for the moon landing at GE! A college Professor himself, I guess my genes are inherited by my Mother's side of the family! Don't tell anyone, as I don't want my Irish Father awakened from his long-deserved rest! Dr. Charlie I bet could write a fascinating book also. A WWII officer who spent his entire career at GE and remarried in his 90's in the Caribbean island to a Boston Irish widow who cooked me a great meal during my last visit until he moved to upstate NY. GE Valley Forge is still there, but now is Lockheed Martin and still cranking out the NASA spacecraft and being monitored by my old friend Jim Suraci who worked on many NASA projects with me at GSFC.

In discussing my other family cousins with Charlie, I told him of Cousin Tom a two-star Navy Admiral and a Deputy Surgeon General of the USA! He replied oh that's nothing, "we have a four-star General in Italy who ran the Tank Corps!" RHIP (rank has its privileges), I replied!

I recall asking Cousin Tom how he got to be an Admiral rank with so many minorities both male and female in contention for that rank. He replied immediately by saying someone needs to do the work! So much for our DC Beltway political goings on. I told Tom that before I retired from the Beltway rat race several females were promoted to new top Directorate levels positions causing many highly qualified and seasoned civil servants to complain about reverse discrimination taking place right here in River City! This action was taking place all over the Beltway in fact all over the USA in Federal Government positions and with DOD and government contractor businesses!

I too sympathized with other highly qualified personnel by hitting the so-called glass ceiling that now not only included race but gender, age whose career ladders were now taken be even green card holders some of whom had names that even I could not pronounce or spell but had a dash on their mahogany row nameplates! Back in the day, only highly qualified people with proven track records were considered for supervisory or management positions. Today, that is a minor consideration on this old-world requirement just as college entry quotas and selection criteria sometimes block most of our nation's best! Equal opportunity is just that and all companies today have written master plans for government compliance to our Fed Gov laws, rules and regulations! I was asked just the other day by my son if I believed in evolution. I replied, yes, however, I think it stopped with my Father's "Greatest Generation"! And from then on, we have seen regression, not progression, this included both he and I as I closed out the conversation!

Individual Contributor: Time to de-compress after 40 years in Management! As a working manager, as my contractor boss described my job, my first 10 years at Goddard saw dramatic cultural and managerial changes all over the Federal government. This also happened with beltway contractors. Affirmative Action, Equal Employment Opportunity, Liberation movements, all collided like a snowplow and pushed the envelope in pushing out old school WASP males from acquiring the higher-level management positions both in and out of government. Legal court battles first won in Baltimore at the Social Security Administration that employed almost 40,000 people, were now taken to many other beltway governmental agencies, including Goddard. The good old boy network lost its power…it was now the good new boy and new girl network, and, like college placement quotas, formal workplace

225

written employment plans based on race and gender needed to be submitted and approved by governmental agencies. Our contractor management was all male, mostly Caucasian, since engineering degrees were prerequisites for employment and few minorities had scientific backgrounds until recently. In our old company, a new female CEO was the first female new player to hold that office. This did not last long, however, after it was found that she hired many females who she was suspected of having too close a relationship with. The old, retired, white, male CEO was hired back to fire her and all her close colleagues. Big changes, however, especially in lost bonus money for not having proper minority balance, soon even affected my role. And, at the same time also, the new younger generation of government players, all wanted their friends in similar positions. Nothing like age discrimination, as I was soon hitting 60! That is how I hung up my helmet.

Called in by the NASA customer close to 60 years of age, I was offered a position until retirement, if I would stand aside and give my job to a younger friend of his, one of my Team Leaders. Think it over, he said, and you will not take a hit in pay either regular raises will still occur. How does it feel to get this news…well, it was like a tough punch directly to the gut? And it hurt! That weekend, still single after all these 20 years, I decided I needed a diversion. So off to Gettysburg, PA I went to really find out what really happened there! Well it took 10 trips for me to see the big mural…the struggle the South had, and the defensive position held by the North.

As a native Pennsylvanian, I was dismayed on how little my education covered the Civil War and this famous battle of just a few days. Once before, years ago, I and my entire family spent the weekend meandering around all the statues and monuments. We only got a glimpse of what really went on…but knew the North won and that was it. Once I went to Little Round Top and Big Round Top and discovered the name, Joshua Chamberlain… I was hooked. How did all those down Mainers get to PA? And who was this hero

Medal of Honor winner Chamberlain and what was his claim to fame. I won't spoil the movie or book answer for you…it is a must see and read historical masterpiece. Thanks to Ted Turner for his filming of this turning point to the Civil War movie epic. I even went to Chamberlain's home in Maine on the campus of Bowdoin College, where he was once President. And later, I went to his modest grave where a small stone had his name and military rank of General. He was Maine's Governor three times and Civil War flags are neatly placed for all to see inside the Capital Building lobby in Augusta.

The second must see site was General and President Dwight D. Eisenhower's farm adjacent to the main Battlefield. As a kid, I watched on a black and white, snowy TV when then Gerald Ford, Congressman of Michigan nominated IKE for President. Later, as President himself, he would sign the end of the Viet Nam War in 1975. Ike won overwhelmingly defeating Adlai E. Stevenson, a Princeton graduate. "I like IKE!" could be heard, over and over, at that first televised Presidential Convention.   In my estimation, the better candidate won and kept us safe for the entire decade after Korea.

Somewhere in between those ten Gettysburg trips, I acquiesced and gracefully complied with my NASA Bosses wishes. Gettysburg for me was my way of coping with disappointment. Like any leader, dis-mounting gracefully, from that lead position is very difficult! I could see how General Lee felt after the July 1863 battle. I could feel what Vince Lombardi felt giving up his famous Green Bay Packard's. I understood how Mantle and Mays felt not facing those 95 miles per hour speedballs and hitting it again out of the park! And of course, that famous scene in the movie Patton when at the end, he said, "fame is fleeting" even at the highest levels of competition.

Welcome to Mission Operations I was told, a new Directorate function for Flight Assurance Management… My journey, in the pits again, just started with this new activity. I needed to revisit functions that I worked years ago and get up to speed fast and make some value-added contributions. Just like riding a bike to me, one never forgets! The more we reported to NASA Management the more they wanted. I was told, and this is no pulling your leg, to make some computer graphics reports on all spacecraft on orbit and then display their weekly performance status. What would you like I replied? When we see it, we will tell you! How about the first one in two weeks! I used the old GE Jet engine plant stoplight (red, yellow and green balls status) technique and they loved it! This was the new breed of management that I faced until finally packing it in. They had the book smarts, however, and with little or no mentoring or real management experience were lost in the woods, without a compass. Internationally, this need for training opened the door for gurus like Steven Covey (soon to be multi-millionaire, who saw the same things in Academia). His great seminars of the top 10 Secrets of Highly Successful People were the latest rage. I attended several and tried to get all at my workplace interested in looking at the new paradigm of workplace dis-organization. Some uninformed folks were even preaching; if the system is not broken, then break it! Remove the old, obsolete organization smokestacks was the new mantra! Insanity was taking over and continues today inside and outside the beltways of America! And CEOs are making record paychecks...some $30, $40, $50 million per year trying to fix broke organizations that they leave worse off than before they got there! Look at HP, I counted Billions, not millions, but Billions of $$ wasted away on such new wave managers! And after failure, they exercised a golden parachute at a young age and live a life of leisure on deferred income payments, while back at the ranch everyone suffers.

Dealing with systems failures and anomalies are part of every engineer's tool boxes. I studied them for all my 50-year career with prevention the name of the game; a quest of every endeavor whether in space, DOD or commercial business sectors. Remember these key words that you can insure risk, but you cannot insure uncertainty! And uncertainty is all around us; accidents happen just ask insurance companies. What was interesting to me was that just until recent times, little knowledge and breakthroughs were just scratching the surface. Many incidents are caused by humans or as we called them operator or cockpit errors. For example, in airline crashes the pilot is one of the major contributors of these horrific events. And the more skilled the pilot the higher the risk of an incident occurring. This is due to checklist oversights that are either skipped or overlooked prior to takeoff or in NASA's case launch or liftoff!

I discovered after studying the subject that oftentimes there is more than one factor that results in an incident. Many times, there are one or more coincidental causes to failures. Therefore, the use of checklists or pre-testing or monitoring events are important. For example, you would not think of driving long distances in remote areas unless you had a full tank of gas. Yet on one Long Island airline crash the crew did just that on a long flight from South America and they missed the runway by only a few miles while running the airplane on empty! How many times have you taken your eyes off the road only to be surprised by on-coming obstacles or even pedestrians in your path and just narrowly missed a crash? We are all humans and mistakes happen!

Looking at the root causes or culprits as we called them, bugs for computer gurus, often shows a series of pre-incident conditions that finally result in a disaster. In our spacecraft missions, forces act coincidently whether it be operator induced actions or sub-system failure that results in a higher order function to also fail sometimes in a domino-like falling fashion. Thus, the need for critical function redundancy in the design stages be it mechanical or electrical in nature. In cars, for example, we have emergency brakes and now even rear-view cameras. In our homes, we have electrical circuit breakers or even local receptacle resets. As engineers, we are always looking for worse-case backups to allow devices to continue to function even after a major or minor malfunction occurs.

229

The human/machine interface today is of such complexity that before an anomaly incident is witnessed by the operator it is too late to react or intervene in the outcome. Therefore, backup systems are mandatory to deal with such situations. One cannot have enough barriers that prevent man flight missions to begin and end without failure. Only cost considerations of a prohibitive level should be the constraint in future systems design and development. Mission success relies on each overseer to use past lessons learned measures to prevent mishaps from occurring.

World-attended Reliability Engineering Symposiums, held at key USA resort locations, were always great to attend. Just going there to listen to the JPL engineers dazzle everyone was a special treat. They always put on a spectacular show with presentations that held us all in awe. This USA think tank is a treasured asset to our country. The JPL folks were always a learning experience for me and all who were lucky to attend. In my last year, I was fortunate to attend my last one in Las Vegas with several other colleagues. My new company boss also attended, and we got along quite well as his experience and background had reliability improvement projects like I did too!

Photo 18.1: Atlantis on-orbit workhorse of the Shuttle Fleet! Also, Man Maneuvering Unit (MMU) space walker, un-tethered, looking from behind it!

Spring Break KSC Seminars – a special gathering of all NASA management was held annually at the Cape, now KSC. This conference gave everyone a chance to get on the same page as to various initiatives going on across all NASA Centers. After Challenger, NASA got every serious about fixing the system so that future incidents could be prevented. That is until Columbia when the so-called shit hit the fan!!! Schedule became a driver of the past missions, now Safety was the new driver and all the culture was now focused on that! After a Presidential hearing and report was issued, new marching orders became SOP. If you would like to read about the two disasters WWW.NASA.GOV is the place, I was fascinated by both and used words as weapons whenever I got static about improving the process. The photos tell the stories themselves, as they say a picture is worth a thousand words.

Now I will give you a glimpse of all the learning that I received to keep up with a career that demanded annual schooling in order to maintain my skills and technology knowledge. In this business, you either keep current or fall by the wayside as many of my generation did avoiding the use of computer technology. If you were out of the business for a few years, you became obsolete and would never survive the explosive waterfall of new workplace tools.... especially the software that was becoming the dominant design driver of the 21st Century products and spacecraft. My post year training at NASA.... ranged from personal computers to an AMA Listening Course to super, high tech symposiums and spacecraft seminars. Let's review a few of them now.

NASA GSFC Flight Assurance Manager's (FAM) Training Course Volume I & II

This was my second assignment upon arrival at Goddard. The first part covered the basics of Quality Control and Assurance. From the Western Electric studies to the Japanese post WWII Deming taught lessons on Quality we gave an in-depth understanding of the principles and practices that were necessary to accomplish what NASA called Mission Success. Part two used all the FAM Task leaders' knowledge of how their individual programs implemented the many NASA Standards and requirements that were embedded into every mission's program plan. In addition, the spacecraft technology was covered inn depth. And each mission's on-orbit goals and objectives were communicated by the NASA Goddard Space Flight Center Project's individual team leaders. Two large volumes of training, study and reference material and instructor overhead slides were prepared for this highly successful training program. It was repeated for several years after, however, as with many classics and with new management, this too, was tossed by the wayside for the new flavor of the month training fad.

NASA Jet Propulsion Laboratory – Pasadena, CA (jpl.nasa.gov)

232

Jet Propulsion Laboratory's at the University of California - Of all the scientists and engineers and even academia gurus that I met during my NASA years, none were more impressive than the superstar thinkers at JPL. During every annual Reliability Symposiums that I attended, were the leading technical institutions and experts of the world, who all made impressive presentations. Everyone was second to the usually last presenter, JPL. And using, their state-of-the-art computer tools and techniques, they dazzled all in attendance with innovative technical presentations on new techniques that we all tried to adopt and use.

Unisys Management Institute was another in-house training program that all high potential managers were selected to attend. It was a review of basic management principles that I was exposed to years ago and even taught to my many college classes. And at the time, another popular corporate program was on Quality Assurance that was the flavor of the month and cure-all for miss-management in USA corporate minds. It always amazed me that people who never worked in the occupation suddenly were made experts and champions of a subject they knew little about. It appeared much like lip-syncing music to me. My new Contractor boss had reached his Peter Principle level several steps below and reflected what was wrong with the American workplace then. Somehow, and some top management never gets this, putting people like him in high places is just asking for trouble with the troops. In this case, it was the old good cop; bad cop roles that I first saw practiced in Long Island, New York years ago. He was despised by all our NASA customers and eventually resulted in our losing our contract that was held for over 35 years! Unfortunately, the course was being offered too late for subject matter such as How to Motivate Employees, and Maslow's Theory of Hierarchy to be taught and be implemented.

Martin Marietta Management Institute was an in-house, corporate level sponsored training course for all chosen leaders in management to attend. It was not opened for anyone, unless you were sponsored by one's upper level manager. I was proud to be selected to go to Atlanta or as they called it Hotlanta for the extensive executive training. Coincidently, one of my MBA students from NH taught the class. He too, moved to the DC area in search of better job opportunities. So, there was little for me to learn, in fact, I could teach the course instead of attending. I was amazed at growth of Atlanta helped by Carter's Presidency. It was now the number one airport in the Southland and was being reborn as a leading business center of the USA. I got to meet a few good friends from all other Martin operations, especially Orlando. I was also delighted to see that Martin was training their managers in the basics of the management. In my estimation, they were lacking in developing a comprehensive management succession program as the good old Chesapeake boy duck hunters were still in charge of those matters. However, as in all situations the real top-level brass seldom attended such training and they are the people who, in my mind, needed it the most!

19. My Special Assignments: Standard Operating Procedure (SOP) at NASA!

John Glenn Research Center (GRC), Cleveland, Ohio - "Cleveland is not that bad a town!" I was told by my new Contractor boss, a retired double-dipper, who delegated everything, including my annual reviews!

The International Space Station Freedom, I finally got to work on it after years of dreaming about it! At last, a future additional in-space Laboratory called the Fluids & Combustion Laboratory Facility (Destiny Laboratory, Combustion Integrated Rack) that is now on board, in place, on-orbit and working!

Photo 19.0: ISS Fluids & Combustion Facility built by John Glenn Research Center (GRC)

Photo 19.1: International Space Station (ISS) with USA Laboratory Module Destiny on-board

As I was saying, in between work assignments, I was called into my contractor's boss's office with this greeting..."Cleveland is not that bad a town!" Never volunteer, was an US ARMY lesson that I practiced in situations like this! Many a MIA volunteer soon found out it was their final assignment! Not that I in my early years did take on many difficult assignments by this time in my career I was a little gun shy! "What's up," I said? Well, he replied, NASA HQ is looking for some people like you to help them in Cleveland. My mind suddenly flashed back a few years. Passing through Cleveland and remembering one of my favorite places in Dayton, the famous Pine Club, and besides, the NFL Football Hall of Fame, I began to listen further. Glenn Research Center is working on a new ISS Laboratory that the new PM (Project Manager) would like a new set of eyes to make sure it's good to fly. This was a tactic NASA used many times, a sort of "second look" they called it. Two heads are better than one was a very successful management practice. For example, in winning proposal efforts at all competitors' offices there were RED and Gold Teams that reviewed the final work to see if it "answered the mail" (requirements).

Several colleagues were listed for me to select as the "Leader" of the new Team. It was a quick-reaction effort targeted for completion in a budgeted 30 days…using my Gov't Experience Rule of Thumb… 90 to 120 days Actual…I always used a at least three times multiplier in any government estimate for getting things done! I knew what my Team could do it, however, most times it was other "uncontrollable factors" that held up progress…sometimes as simple as a Fed Gov't signature was considered a "stopper"!

I accepted on one condition that was I be able to stay at the Marriott Hotel in Cleveland, a place where we could get a decent meal up to midnight, for I knew we would be working around the clock to get the job done. My NASA colleagues, several of top performers that worked on my team before on difficult assignments accompanied me to Cleveland. We assessed the engineering analysis they requested and developed a schedule to complete the job ahead of schedule as always, our practice. This gave us a window of time to use if needed to meet the actual deadline. Essentially GRC wanted to know if the Laboratory met all the many NASA Engineering Requirements. Several Program managers were assigned over the years and the latest one did not want to suffer being a victim in case of failure. Very smart guy, I thought! For he could always blame us, if things did not work. Our analysis uncovered several gaps and all deficiencies noted could be fixed prior to the official launch date. Toward the end of our work several complained that it was an impossible job to complete. However, I recall on many assignments after working many days and nights this was always the complaint. And as I recalled General Patton's famous quote he used in the battle of Sicily, "L'audace, l'audace, toujours l'audace". I used the same tactic to convince those doubters to finally complete our assignment. GRC was delighted with our efforts and once on board the ISS the laboratory performed as designed.

Mission Operations – Another round robin tour…. KSC to White Sands…the NASA Tracking Stations.

I was told that no one from GSFC ever did an audit of the NASA Tracking Stations in the field….one of my many "Special Assignments"! The Good Old Boy CS network knew how to cover their bases. Where the hell is Ponce Inlet, Florida, I asked my Govies boss at the time? "How the hell should I know," he truthfully replied? No BS from him!! Find out, go there…let's see all the NASA Tracking Stations Operations. Well, I did, and Ponce was in a remote area, just north of the KSC base. Closely associated to its sister-tracking site at Merritt Island, we visited the latter and then the former in a two-day window. I felt something strange about the close-mouthed contractor troops who were our guides on our many visits. After getting to know one of them and said we were not here to cause trouble or hurt anyone, I asked at lunch one day what's going on here…why the QT? A nice straight shooter guy, as all of them were to us, he said a layoff just occurred and our pay was being cut along with our benefits. No union had made in roads south of the border he continued to say, but they wish they had one now, he replied. I did not see their big picture plight until I arrived at White Sands, NM. While touring, I noticed a classroom that looked like new employees…all from Mexico…. being taught English. What's that all about, I asked? Oh, just some replacements being trained…a lot of the folks here are retiring so we are training their replacements.

I looked at my CS boss and fellow audit partner and he looked at me with a disconcerting expression. What was happening here with this new small business contractor we both thought? Later, we discovered this practice was going on in many places with the new small business set aside contracts open only to small company players. And new players were coming out of the woodwork to take over these work assignments. Teams that were around since the moon

238

landing days were disappearing to as John Glenn said; his biggest fear was the contract went to the lowest bidder. Now usually, a minority owned small business owned by Indian Americans (owners whose roots are from India) and even American Indian and American Eskimos, all are now sharing in the formerly closed clubs of Corporate American owned business sandboxes.

For example, while I worked for several of these same minority owned businesses two of them sold out with each owner making over $200,000,000 in selling their companies. In total, just about when I retired, I counted almost one billion dollars were made by selling several minority-owned businesses just from the local Greenbelt, MD area. Not bad for just a few years of developing a Fed Gov contractor business.

Full Circle: One More time: Penn State University (PSU)…my youth desired, but unfulfilled alma mater!

One of my last places to perform a Flight Assurance Mission Operations Audit was in PA at Penn State University's NASA Operations Center. Nothing like going full circle during one's career! Started here in 1957 and returned in 2008….50 years later…WOW! WOW! What a ride, if I must say so myself! Things were quite different this time; however, I had the decision-making role of pass or fail and assigning a grade A, B, C, D, or even an F!

Recalling my Graduate School teaching days when everyone needed an A to get their companies to pay 100% of their tuition. And one reason I stopped teaching due to serious grade-flation problems especially with foreign students! Who were having English language problems and when I complained was told, you're not teaching English…I replied, "You are going to send these people into the work force, and they cannot even construct a sentence?" After 10 years, I decided that was enough…it was no longer a hobby to me anymore and I did not even need the money, since my day manager's job salary was now enough for me to live on. I never looked back and those years were a pleasure to do!

And I learned the MBA course subjects enough to challenge new candidates for a long time. As we read back on earlier pages, I identified all my course descriptions and titles. I started them all from scratch, using only my student notes from classes taken at both Northeastern University and Babson College…two great Boston 128 area institutions that I had the pleasure and luck to attend and graduate from!

20. The New Culture and Writing on the Wall for me to retire…. No Job lasts forever!

As I walked around NASA's mahogany row offices in 2008, I could not believe the names on the doors…I had trouble pronouncing many of them and they a lot had dashes between their last names. Gone were the Tom, Dick and Harry's people of my generation who had made it to the top of their careers. I began to see big changes 20 years earlier, in my Graduate student population cultures. Many of them came from India, intelligent, book smart and well read. However, social interaction with other ethnic students proved difficult in those days. It reminded me of my Grandparents always talking in Italian in the household, however, whenever I used Italian, they told me to speak English! Visiting Italy years later, I picked up using Italian after only a few days. In fact, on my last visit, I almost wanted to remain there…. what was I doing on the other side of the planet, this is really where I belong…were my innermost thoughts?

For example, on my West Coast business trips to California University NASA partners, I was amazed to see at lunchtime social groups of the different cultures not mingling together…Asians were on one side, African Americans on the other. And it appeared that these students were the new majority in LA and San Diego colleges. Yes, America over the past 50 years has changed right before my very eyes.

More Mission Operations Systems Audits…

Johns Hopkins Applied Physics Laboratory (JH-APL) – I recall as a kid seeing then President's Eisenhower's brother as heading up this icon of American Laboratories. Many NASA Goddard folks came from here since the WWII; APL had a Navy Laboratory that was the roots of new technology for the military. In fact, one of NASA's recent Head Administration was a Hopkins APL Director prior to running NASA HQ. APL had a NASA contract at the time and was viewed as a larger player on the near horizon. I was part of an ISO9000 Audit team responsible of assessing their

Mission Operations and Spacecraft Tracking Operations. Their Director later became head of Goddard and in my estimation, was well qualified for the task with his business mindset. I was impressed by his welcome to APL introduction upon our team arrival, His prior career was in industry and what all these college and university laboratories need is more managers like him.

After several days of working closely with his APL Team of managers and reviewing their in-house practices and documentation I scored APL highly for an outstanding operation. They worked hard to make sure things were done right and lived up to their high-quality image as an USA icon!

And back now to **Penn State University at State College, PA**, where I began my career studies. Years later, it has become a leader in training some of our best engineers and managers in the high technology business. A saying inside the DC Beltway is that Penn Stater's occupy many levels of management below CEO and is reflected by numerous alumni at VP and Director level jobs, but few at the highest! My take on that is they are too important workers to elevate to those highest positions…as one graduate said to me, someone must do the work! My first cousin two-star Navy Admiral Tom said the same thing when I teased and asked him why he did not go higher to three or four stars… somebody need to do the work! Yes, you are so right, I said…been there done that!

While I was working at Penn State, I visited the college bookstore and picked up the latest Statistics Course textbook that used Microsoft EXCEL as a tool. It had a DVD inside the rear cover and was written in an easy to use format. I found out that all students needed to take this course and was impressed by that requirement. Penn State is a NASA and DOD Project house; I was sent to look at the NASA side of their operation. With all the post 911-security measures, it was not easy to find this operation. However, they did occupy a separate building and small signage was on the front door. What struck me, as did all of NASA was the number of foreigners employed here was many UK engineers all experts in their various specialties. It was not APL, however, and a much smaller group of close-knit personnel ran the place.

White Sands, New Mexico Tracking Station had been on Goddard's list to audit for years...politics being what it is however they dodged that exercise with all kinds of excuses. My new NASA boss, actually old client, promoted up the chain of command was determined for us to visit this place. I discovered why upon my arrival to the site. We were the very first to do an audit of the tracking station that had the same folks running it for years. All seemed in order the place had powerful technology well maintained. But my eyes were trained to see through the smoke and mirrors and dog and pony shows often accompanied by such site visits. I discovered a training room with about 50 students being taught...you will never guess it...not science or technical subjects, but English! All students appeared to me to be from south of the border...the contractor was training the next generation of support personnel...the same situation I discovered in Florida. No wonder no grey-haired folks manned the place...either they retired or were replaced for lower wage workers. I don't think it was for me to discover this situation but, after 30 plus years and more importantly, lots of street smarts, I was not the normal ticket puncher passing thorough with rose-colored glasses. The Resident Government Representative in charge did not want me to make my discovery part of my report. However, I did as that was what they were paying me for, tell us what is really going on out there, is what I was instructed.

Merritt Island, Florida (MILA) was a critical tracking station for all shuttle launches and a well-run operation with seasoned players some for their entire careers. Located off site from KSC, MILA as it was called, had the latest equipment money could buy. Again, this contractor was on board from the beginning of the space program. They even tried to sell me a former test engineering station expert, with new upgrades automation (now they did have my attention being an automation admirer back in the day!). And what they said was computerization of the entire operation. I wondered who approved this so called "improvement and upgrade?" What I saw was those taking manual operations and volumes of instructions and placing them on CRT screens! Not quite either automation or computerization in my world, but millions of dollars went into this upgrade. So much for incumbent contractors doing their thing. I could see it now once on CRTs and in computer memory this information was theirs to keep…trying operating without the old system!

In 2007, I was given an opportunity to address the annual NASA Supplier Chain Conference held at Goddard space Flight Center. As we were approaching the 50th Anniversary of NASA and my 50 years in high technology, I wanted to provide all attendees and on Goddard's home page web-site an archival resource for what I thought were lessons learned that could help all attending world-wide companies. Also, I discovered a great systems analysis tool developed by a MD and PHD who worked for years in DC governmental jobs. His tool was called "Barrier Analysis"! My having spent several years analyzing NASA On-Orbit spacecraft and world-wide tracking station anomalies and failures, I also took a shot at how these anomalies reduced and even eliminated in future design and development efforts. However, these techniques are applicable to any product and have universal engineering application. Therefore, I am adding this to my book for as they say "an ounce of prevention is worth a pound of cure as was the standard back in the day! See my Appendix for more details of these important measures.

21. Not the First Time, The Second Time, but the Third Time Really Means Retirement!

Coming to the end words wise…but, more to come…. I'm not finished patient reader!

Many readers by now may say, "This guy is a masochist, why continue playing the management game!" What is the payoff, the payback, the old business measure called return on investment (ROI)? One answer and reason are that you become locked into the system and it is very difficult to pack it in! Another is that this high technology quest is addictive almost to the point of being an obsession! Ask a baseball professional why he continues even after his peak season is behind him. Or a Doctor, a Surgeon, whose job specializes in performing operations successfully like no other one can. Once the money part leaves you as a driver, one seeks more challenges…higher mountains, bigger stakes! Or an actor, who has done 20 years and still wants to reach for the magic brass merry go round ring as he continues his life in different role after role to reach his/her goal. And never forget our roots or how far we have come to escape the boring, repetitive, mundane work we saw in the past world history. And how we promised ourselves to do better, no matter where or how or what sacrifices we needed to make! And we can often today look back, far away from those unpleasant, back in the day, almost forgotten memories. Maslow's Theory of Hierarchy certainly holds true, once we are satisfied with lower fulfillments, the higher ones' kick in…recognition and achievement tantalizes us all!

If anything, my end goal was set too low (I first only was wanting an Associate's Degree, AA), I under-estimated who I was, and who I now became even without a degree. I wonder if Gates, Jobs and Ellison had the same experience. For I had achieved far more than they all did in academia world, but I made significant tradeoffs in future entrepreneurship endeavors. My generation was always haunted by Depression Era parent's upbringing! Our parents and their generation were afraid to take chances and saw the downside realities of their struggle, but never the upside potential of being your own boss and starting your own business. They were delighted if I could just get a job with the local Bell Telephone Company, in any capacity, and stay happily employed your entire work life! As I looked around at other so-called managers in the business, I had more education, hands-on experience and now even the necessary street smarts to surpass all of them up to the CEO level that I was originally trained to achieve. However, I never was willing to "sell my soul" or take the leap of doing my own thing to achieve that last few steps up the corporate ladder. My graduate school business thesis entitled <u>Small Business Management Assurance</u> (SBMA) that disappeared and was never returned to me. I after reading almost every business publication written at the time that forecasted the new IRA 401K, 403B, retirement investments that today are billion-dollar business organizations like the Fidelity Fund, Janus Fund, Edward Jones, etc.   At 33 years old, with three kids, several horses and countryside estate, it was hard to give up all my comfort factors, little did I know then that I would soon lose them anyway, through a No-Fault NH divorce! Hindsight is certainly 20/20!

My future path.... Retirement 101 and completing my lifelong travel bucket list!

During my work years the one place I could ever seem to be able to relax was in Florida. Whether it was the weather or laid-back lifestyle, I always breathed easier there whether at Winter Park where I lived for almost two years or the beaches, New Smyrna, Cocoa Beach and in winter, Clearwater Beach. Another place I visited every winter in January from 1985 on while working up North in Baltimore, was to Mexico, specifically Cancun and Cozumel! I loved them both, but the less traveled Cozumel became my favorite. The underwater museum at the public beach, the great 4 Star restaurants and occasional Churchill Cuban cigar was hard to match anywhere on the planet! And the people and delightful village are right out of Hemmingway's Key West Era.

Prior to packing it in, I decided to go somewhere in the World that I always feared...the Orient...with Korea and Viet Nam in the back of my mind...not very wondrous days for my generation to live through. During Korea, from 1950 to 1953, we were still undergoing Government rationing and missing our local military neighbors, people who I knew growing up. One poor soul was our local soda jerker who made the most delicious ice cream sundaes at Borr and Casey's Drugstore near my parents' home. I still remember his smile visiting his counter...How I can help you he asked.... Black and White Sundae, I replied, on my days when Dad would pay me for helping him do a big wiring job! MIA the headlines read about his disappearance in the police-time action. Wow, I said, I knew that boy...now in a grown-up man's war and now it's innocent victim.

Using American Express, before they downsized in Europe and elsewhere, I sat down in Baltimore and discussed a China trip with a gracious lady getting ready to retire who helped me greatly! She had me book a round trip, non-stop flight with Cathay Pacific on a Boeing 747 from NYC to Hong Kong, China. And, I discovered a new planet, like Marco Polo did. Landing at Hong Kong's amazing, large, new airport, all I needed to do was get on a fast-express train right to my hotel in the Central Business District. My bags were already in my room at the hotel...before I got there. Hong Kong is one of my favorite places on the planet. The British need to be honored by what they did to bring that place to the 21st Century. Hong Kong is unique, a world-class destination. If you go anywhere in your life go there, you will not regret it. We Americans are ignorant of the Orient and its renowned culture. For example, in Rome, Italy, I saw 2,000 years of civilization, but in China, 20,000! China ruled the World four times, I discovered.

I love to tell this story to my American friends. In the Middle Ages, the Pope sent hundreds of priests there to convert the Chinese...however, none returned to Italy! I think they were converted in reverse. Like the Italians and most Europeans, family culture is everything there, as I saw growing up in my hometown in PA. I have a deep respect for people whose values are like mine and the WWII kids' generation. Our traditions and priorities of "family first" was and is a powerful cultural value. And it was displayed in every part of the Orient that I was exposed to. I thought, those folks are just like us, the same needs; food, clothing and shelter and family survival. They also disagreed with their leaders like we in the USA did…we love our country and trust, but verify all others including your own leaders… should also be our charge!

Hong Kong is a business town like NYC, young people well-dressed, yes, better than in the USA! Classy wardrobes, like we had in the 1950s, make up a large part of the population's desires. My first day, I went by bus to Stanleyville, a British upscale suburb and dined on German food in a former German barracks converted into a Bavarian styled eating place. Later, I dined on the Big Red Boat harbored in the nearby waterway surrounding the entire island area. A must see on your trip there, is to go to the top of The Peak, during a clear night, overlooking the cosmopolitan city. This is a sight to behold, nothing like looking down on a miniature, Christmas tree train-like display, only real! Marvelous skyscrapers tower for miles and miles enhanced by the unique lighting displays. Nowhere on Earth, even the Big Apple, is there such a wondrous sight to behold.

From Hong Kong, one takes a train into the mainland connecting Shenzhen, a new city devoted to business and manufacturing. As I took the train going through barb wired, highly policed presence passageways, I never in my life thought I would be doing this. Entering a communist country, I flashed-back to my comic book era when the war year stories biased my thinking that Asia was a feared place for Americans like me! A chill ran through me and I hoped I would return to Hong Kong unharmed from this once in a lifetime trek. After arriving in Shenzhen, however, my fears were completely unfounded. After filling out some forms for customs, I joined many foreign travelers and it seemed like the west coast ports of entry to me. My hotel was taken right out of Broadway, as I entered a long winding staircase appeared with a tall, statuesque Asian Girl dressed like a Hollywood movie set singing American blues tunes. It was truly a Shangri-La Hotel, one of China's finest chains. Everyone was enjoying a free show with many tables and cocktails served as well. I gave it a 10! Oh yes, during the night my room became uncomfortable. Calling down the front desk, I was told that the government had diverted electricity to the industry sector there was nothing they could do about it! Checking out the next morning, I was given a considerable discount for my discomfort, my only complaint. To try to describe China to someone who has ever been there I just say multiply big cities and buildings in the USA by at least 4 times larger and you get the scale of China. 1.3 billion People vs our 350 million, about a factor of four, is how I derived my multiplier. Returning to NYC was like seeing a country at Epcot in Disney World. The USA is not that big a Country, I thought. From Shenzhen, I visited Zhuzhou via air from Guangdong Airport via China Airlines. All their jet airliner planes are made by Boeing, Seattle, WA, USA. I recall the President of China visiting Boeing a while back giving a 1,500-plane order to them and receiving a standing ovation by the many grateful employees of Seattle.

On China TV their Fox Business channel, I saw Fed Express President Fred Smith (one success story I used in my Grad School classes) was on that weekend and he said that in ten years, FED EX China would be bigger than all the other countries combined. They too, ordered 1,000 new jets and he announced the opening of 1,500 FED EX offices in China! How's that for championing entrepreneurship! And recall today Apple has over one million people working in China making their phones, iPad, iPod, and soon to be …I Watches!

Arriving at my destination, I was delighted to see an original Chinese City intact from its origin. My day started at a Chinese coffee house going in for an American breakfast. I was shocked to see down the long main street, employees out in front of their businesses at almost attention getting their daily instructions from business bosses. Something like Walmart does in the USA but inside their building. My First day was to include a tour of a Chinese Emperor's burial site…some 15,000 years old. And a large museum dedicated to their Korean War soldiers. I could not handle that one and felt very uncomfortable during our fast walkthrough.

Retiring at age 68 to me was my final work attainment. And playing in the big leagues with the best of the best was reward enough after 50 years in the high technology business. So, what does one do after pulling the plug on work that always remained challenging? My short answer keep traveling and enjoy the time in great places with the best of friends.

One of the first trips I made was to Ireland with my oldest son Chuck Junior. Chuck, Jr. is an Amherst University and Michigan Law School graduate and member of NY Bar. He was a lucky that I could afford to send him to Lawrence Academy in Groton, MA. for high school, only a dream for me when I was a teenager! I got on a commuter plane from Philly directly to London in the summer of 2009. It was after my 16-hour flight to Hong Kong, a short hop of just over 5 hours. Once in the UK, I was fortunate to be a guest in the President's House at Oxford University. And, for several days, prior to our flight to Dublin my son Chuck, Jr. even stayed in Bill Clinton's room! During each day tourist would gather outside and take photos where the future President of the United States lived during the Vietnam War! As he like tens of thousands of others avoided the USA military draft. As I recall some 35,000 others escaped to Canada, later to be welcomed back home, to the USA, by Jimmy Carter after the war ended. Oxford is quite a nice graceful community and an international one. Students from around the world attend the more than 20 colleges that make up the academic community. They long for Fellowships there were they can enjoy the serene atmosphere and highly civilized lifestyle.

On Sunday both my son and I attended grand Church of Christ that is a must-see attraction.

Photo 21.0: Winston Churchill's Birthplace at Blenheim Palace in Oxford shire, England, UK

Everyone thought I was visiting professor on sabbatical or conducting research at the magnificent library surrounding the town. We also got a chance to visit were Sir Winston Churchill was born in a grand Castle estate with manicured and highly developed landscaping surrounding the estate. The Oxford weather was just wonderful when compared to our week in Ireland. For every day there, for seven consecutive days, it poured poor cats and dogs as they say. We went through several umbrellas that needed to be discarded and my son resorted to buying an expensive raincoat, with a hood, that he wore every day we were there.

Photo 21.1: Street Scene in Ireland...home of my Irish ancestors long ago!

We spent two nights at the famous Imperial Hotel in York were one of the Irish Republic founders stayed on his last night before he was assassinated. I did acquire a taste for Guinness since it was much stronger and tastier than what we get the United States.

Upon our final arrival in Dublin, after a sold out U2 Concert weekend, we were welcomed by our gracious bed and breakfast host. I couldn't help to take in the fireplace flames roaring in the month of June. Such is the climate of the Emerald Island. Oh, Mr. O'Boyle, he said, "Welcome Home!" and gave me a big hug! He continued your Grandfather was from the West Coast and your Grandmother came from County Mayo, it is an honor to meet you! Wow was I impressed by his introduction. My son said that Henry Kissinger had stayed here several times and the place was highly rated. "Make sure you come down early for Irish breakfast," our host said. "We have a different menu every day and I think you and your son will like it." And we did, it was delightful every day.

Our plan was to take the Irish train to several cities inland on the West Coast. And end up at Shannon Airport for a flight back to London and then for me, back to Philadelphia. I was told that long ago; a train ran up and down the West Coast of Ireland a sight to behold. Today, however, one must rent the car or join a tourist travel group to make the same trip. We visited all the tourist places and I certainly was impressed by the Irish people. And we loved the beautiful green rolling hills that seem to cover all of Ireland. We arrived at our last city near Shannon Airport and entered a new, ultra-modern hotel. Upon signing in at the front desk, we were asked if we were with the Dell Company upon signing in. What Dell, I asked? Why the computer people, they replied. No, I said no Dell folks here! Okay they said now you can register as the Dell people are persona non-grata here and will be turned away. Later that night at the bar, we got the inside skinny. The Dell Company had a large factory nearby and just laid-off thousands of Irish workers and moved their entire operation to Poland, after tax advantages ran out in Ireland. This hotel was built to house all those visitors and today only a handful of customers including few new visitors had registered. Their days were surely numbered, since it was built for 300 guests with only a dozen or so here today. So much for the recently booming Ireland economy. I told my son don't they know there's a United States recession here since the prices seem to reflect the boom years. Many offices nearby were empty, a prelude for what was to come. Today, Ireland still suffering, however, the Irish people are resilient and as they say this too shall pass. Somehow Michael Dell took advantage of the Ireland tax advantages for several years and when they disappeared so did Dell! On to the other tax-free nations such as the former Iron Curtain countries. Except for the daily rainstorms Ireland is a must visit place for the country's attractions, people and culinary delights and of course, beverages, are all special treats to behold for a lifetime.

253

Retired – many more things to do and so little time to do them!

Remember the...state of Maine. Maine calling... More lobsters please! Since my first visiting South Portland, Maine in the late 1960s when Jimmy DeMillo's original seafood restaurant, where one could buy two big stuffed lobsters for $3.95, Maine is addictive! From Ogunquit to Bar Harbor and Cadillac Mountain, I'm often amazed how many Americans are not familiar with this USA territory. My college classmate good friend Bob Tropea invited me to visit him once we were reunited again in the 1990s. I'd hop a flight from Baltimore to Portsmouth for $50 and have a weekend living life as the Maine motto says, "The way life should be!" Bob was teaching college for years at the University of Maine and their business courses at many of their satellite campuses. To my favorite Maine locations are old Orchard Beach and Bar Harbor at extreme locations by car far distances from each other. For seafood lover like myself, Maine is a place that can satisfy any appetite. And Marginal Way, at Ogunquit, is a one-of-a-kind trail along the Atlantic Ocean that must be seen and walked slowly from one end to the other end. And not forgetting Freeport's LL Bean's grand complex of stores open 24/7 with all the top fashion stores located at the town's downtown center. Maine is a great place in summer and fall; however, weather can get nasty there in winter and spring unless you cannot be a snowbird, one can suffer for months. For me a new snowbird there is nothing like the sunshine state of Florida, where I live for several winter months of the year.

The sunshine state of Florida, a bucket list destination for many in the World, is to be visited and resided in, especially in winter. Central Florida cities like Orlando, Altamont Springs and Maitland are great places, however, my favorite middle of the state village is Winter Park. The first time I visited here in the1980s, I found my home in this great community. However, as I said earlier, after two years first living there, the winters in Central Florida can get quite cold in January with heavy frost occurring even on the East Coast.

Cocoa Beach winters can even be quite unpleasant. My year-round location is the Tampa, St. Petersburg and Clearwater beach region. In wintertime, we folks in Central Florida could be found weekends on the beaches in the Gulf of Mexico. Mine was Clearwater Beach, rated by USA Today as Florida's Best Beach.

From my first visit there in the 1980s, I made a promise to myself, that when I retire, I would live here in the wintertime. And after being here five years now, I did fulfill that dream. In the 1980s my parents would also winter near here. I told my dad than to buy one of these on the beach high-rise condos for $30 to $40,000. He said, "Why buy, we only come here for three months!" so much for the depression era generation. Today those same places go for $300,000 and up! On the beach houses here are over $1 million and nearby housing can go up to several hundred thousands of dollars depending on the area. Along the Gulf of Mexico coastline from Tampa to Sarasota to Venice and Naples, all the way to Miami, shore beach land is now taken and occupied by the as we call them here the rich, the very rich and the filthy rich Americans, Europeans and South Americans and world-wide "folks who made it" in our USA Society. And most of this growth occurred after the 1980s when I first resided myself in the sunshine state.

I recall betting my Sister in 2011, that I could rent a place at the beach for $2,000 a month…well guess who won…she did! Nothing was available near her place, the Sarasota beaches, short term wise for under $3,000 a month. And the future is amazing…. with the boomers, all wanting to have a place down here, big plans are in place to make it bigger and better than ever…. Florida is one place where dreams are made to come true. One example is our favorite Norman Rockwell-like town of Celebrity, a Disney created village that is right out of the movies…Hollywood wise. Again, if you can imagine it, it can be done, and Celebrity is a great example of that saying.

22. LESSONS LEARNED & SOLUTIONS APPLIED!

Most Fed Gov Agencies such as NASA and all the Military services have what is called a Lessons Learned database to archive important events and incidents, especially process related ones so failure or close call history is not repeated. During my time, we recommended and added many lessons to this valuable NASA HQ resource. NASA Program Managers used these resources to "get up to speed" on mission history and mandatory safety issues that go back to the KSC Apollo I capsule fire of January 27, 1967 that took out several heroes. Space.com's excellent web-site has many photos of this first NASA disaster. Three Astronaut Pioneers Lt. Col. Gus Grissom, Lt. Cdr. Roger Chaffee and Lt. Col. Edward Higgins were lost in an on the ground, practice exercise that later helped create new Apollo fire-proof cabin environments.

During my career, no matter what the outcome on any endeavor…Fail/lose or more importantly pass/win, it is extremely important to review how and why success or failure was achieved. In college, I learned of the Pillsbury Management Approach lesson that strategy was superior and most important in accomplishing any endeavor. In sports, an example in football was the invention of the Pass, T-Formation, Single wing, Green Bay Packers Power Sweep and even the Hail Mary; so well demonstrated by Boston's Quarterback Doug Flutie. And as Coach Vince Lombardi emphasized, "Winning is Everything!" Simple to say, however, as was mentioned earlier, it starts with detailed vision and planning…strategy was defined simply as what you are going to do and tactics detail how you are going to do it!

For me it was my own internal driven initiative…. wanting a college degree, but not prepared at first to complete one. And having the resilience to still pursue my dream was an important factor. But also, never giving up and still prodding along in little steps with focus always toward my objective. Taking remedial courses, reading and working side by side with very smart people…world-class experts in their field.

All combined, helped me achieve and even exceed my desires. Emulation is as good as creation I once heard…and I emulated often in difficult situations.

So, as we end our 50-year journey through the good, bad and the ugly times of my career, what lessons learned could be passed on to the next generation of high technology pioneers. For, in my estimation, we are just beginning to uncover phenomenal discoveries in the future, that will make today's technology seem primitive and obsolete as we move forward and see time pass in the rear-view mirror of the 20th Century.

Here are a few of my lessons learned and humble advice to my gracious readers, especially you who are young and are about to begin your own journey of life! And of course, to our next generation of managers! Those at the middle level where most of the tough jobs are. And were one's success can lead to the highest managerial levels in the American or now international business arenas.

First have both short-term and long-term goals and be optimistic make them always be higher than you think, for once opportunity knocks you must seize the day, as they say. Have a plan besides just goals, put it in writing with a timetable for attainment. For example, I probably lost about 10 years waiting for my academic credentials to kick in. I also spent almost 10 years doing the academic circuit just as a sideline aspiration fulfilling but, in many ways, just a steady income stream. That precious time could have been more meaningful used to attain higher-level opportunities.

As it is taught in business ethics classes, it is better to walk away from situations where your reputation image or character can be smeared! When it comes right down to it, your education and reputation are two things they can never take away from you! Mankind's problems often created by men can also be solved by men, that's what separates us from the animals and the beast of burden.

As is in every sport team, especially the winners have what I called secret weapons. Get a cheerleader or partner who can help you achieve your goals. Whether it is getting that extra edge most of the successes that I witnessed always had someone in the background who was supported and influential to one's career some of the highest level of managers have partners who threw the best social activities. Partnerships are hard to beat assets when climbing the corporate ladder! Being just happy to get out of bed every day to do something you love to do is the best kept secret to longevity in my book!

Find out what you are good at doing and strive at being the best! Look at some of the greatest generation entertainers like Sinatra and his so-called "Rat Pack"! Nearly all never went beyond the fourth grade in school, yet they were the best in what they did in their chosen careers. And look at today's leading companies and study their roots. Many dropped out of college to pursue their dreams. They understood that there is nothing like starting at your own goals and taking advantage of their youth's drive and ambition! These traits energy, drive and ambition all work hand in hand in creating the right path in making the right decisions at the right time! In my case, love what your doing was probably the number one driver for me, even before monetary considerations!

And besides a plan, one must have a vision for not only your future, but also the future world that can fit into your vision. This vision should be about what work is satisfying to you and take consideration your personal strengths and weaknesses; for example, being in sales and when it comes down to it everyone is in sales! Not liking working with people is a reality some people need to understand being in business if you like it is not a religious avocation. Trade-offs will be necessary if you're going to survive and thrive in the real world, however, remember to never sell your soul and never cross over to what is not the right path for, as they say, the road to hell is paved with good intentions.

For me during my most trying times, it was my daily run that saved me. For almost 25 years, I exercised several times a week and as my Johns Hopkins Doctor and School of

Medicine Professor said, "I want you to sweat 20 minutes a day to keep yourself healthy and live a long life!"

And as one of my childhood hero's, beloved by most of his player's, Coach Vince Lombardi also said, "Timing is everything"! Whether it is in your planning, investing, or daily work decisions...you can go forward, but few can ever go back. Each day is like writing a book, each month a paragraph or chapter that eventually makes up our own lifetime book! Taking the right path is most of the time what leads us to our own individual successes. Not quite like Yogi Berra said that "if you come to a fork in the road ...take it!"

You can be a big fish in a little pool or a small fish in a big pool! No matter makes contributions using all your skills, experience and training whether it be in computerization, automation or systemization. Leave things better than you found them and that can be your legacy! In my case, learning and using the systems approach was a powerful tool and helped me make large leaps with my ideas instead of little steps in bringing about efficiency and effectiveness improvements. It also paid off in monetary rewards, whether it be the Chairman's Awards that I repeatedly won in several consecutive years or my own personnel evaluations when it came time for raises. I often evaluated myself as I myself practiced with my direct reports...for who knows better than you what contributions you made to your organizations profit and loss basis. Remember always business is there to make a profit and in HP's charge to make contributions to mankind...that they did are still are doing!

Photo 22.0 President Ronald Reagan's Air Force One at his Semi Valley, CA Library

Lastly, find something to do besides work! For me during my most trying times, it was my daily run! For almost 25 years, I exercised several times a week. And during my workday instead of smoking like most of my generation did. I walked and enjoyed the scenery. I always return refreshed and with a clear mind from this daily routine. I started running in my 30s after studying the habits of fellow higher-level managers who were older than I, but still acted young for their age. And having started running in the Boston area, the weather was certainly helpful in continuing my exercise routines. And remember take those vacation times and any time you can away from the office! Easy for me to say now, I know, however, amid manager's daily fire-fighting jobs, it is so difficult to toss aside job pressures. One must practice being able to switch off job thoughts as an escape from the stressful world that is part of one's chosen career. I always reflected on work problems and during these off-work times found the best solutions to my most difficult problems. Conflict management is part of the work world and some people thrive on pressure while others completely collapse. With the proper skills and training and experience, one can deal with such pressures and skillfully take the right paths toward successful conclusions to nearly all problems.

The simplest solution is often the most economical and most elegant! Look at Apple products and even HP and in their day, Sony. Steve Jobs was a stickler for quality and for simple details in Apple products…he learned some of this as I did from working in his case part-time job at HP, in mine it was in my full-time job at HP. HP was like getting an advanced engineering degree with hands on experience at the same time, while working on the latest high technology.

I always loved the underdog, because in my own case, I was one. Competing with older guys from grade and high school and in several sports that I loved. Being 13 years old and competing with 16, 17 and 18-year-old guys can be intimidating. So, I needed to be as street smart as they were. Never one to withdraw from the competition, I used my entire God given powers: spirit, body and mind to succeed and win. These lifetime gifts seem to be lacking today in our leaders, both in business and government. As Dr. Steven Covey so clearly described in his book on the "Secrets of Highly Successful People", when applied, they do work like magic. Success breeds success. There is even a success factor in failure, as the best seller "The Success of Failure" explains. I first saw the book in a founder of Sanders Associates office! I said to myself "wow, here is a high-level entrepreneur reading about failure, he must be a brilliant man" and he was. His named was Mort Goulder, one of the original Sanders Associates and a neighbor in my NH bedroom community, called Hollis.

In studying leadership for my Management class preparation, I recall reading about the serious downside of the role. One thought about the consequences of being the leader in times of trouble. It is not all about the glory days, there is a downside…the so-called "scapegoat syndrome" as it was referred to throughout my entire career. A leader can also be a victim in situations that go bad. Again, in sports the best example is that NFL coached are fired, in business, CEOs turn over quickly, if goals and financial results are not met. No matter what level, a manager must be prepared to deal with situations beyond their control; be it the economy, the environment, disasters even incidents that happened before his watch, any unplanned for event. My Northeastern Marketing course called them controllable and uncontrollable factors…so aptly defined. As I told my students often "you can control and insure the certainties, but you cannot insure the uncertainties."

The higher up the food chain, as today's boomers like to refer to the world, the bigger the defeat…. the "ecstasy of victory, the agony of defeat" so well described in the world of sports.

Looking back from where we are today in 2019, since 1995 (25 years ago), with our computer technology shows what it takes to become a big time, long term player like HP was for almost 50 years. For example, in Microsoft's case, when their products were first developed, they never really took off (in Fed Gov) until DC, inside the Beltway lobbyist, were hired to stop hitting the company with legal obstacles. Finally, Microsoft saw the light and began political contributions to grease the skids inside the crowded political arena. As I said, I witnessed first-hand how DC power influenced our workplaces by forcing us to use first generations of Office in the early 1990s. All correspondence required using Microsoft! I personally liked and used Word Perfect and Lotus in place of Word and Excel. In my estimation, they were superior to Bill Gates' software. And we always had those die-hard Apple users to contend with. Soon however, all Fed Gov proposals required the use of Office and we all grumbled with displeasure. Later, the broad application of Office became much better, however, even as I write this book on the latest version of WORD, somehow, I think we are going backward in user-friendliness and simply ease of use of this powerful software. And, the large new market that was created to help prevent virus from driving you crazy…perhaps another design problem that should have been corrected at the source.

Meanwhile back in the early days at HP and Apple, high-technology gurus were designing all new generations of personal computer products. At Apple, software designers were developing proprietary stand-alone hardware and software with what was termed as closed-end design. This blocked hackers and viruses that in the case of Microsoft (with their open-ended design) created a whole new industry of security protection just to use their products. Even today as I type this on a MacBook Air with greater ease than on my many HP laptops and desktops, I credit Steve Jobs for his foresight in developing products that have a high degree of susceptibility to preventing internet threats. He made the right call to develop his own software, in house. HP missed the mark as its hardware dominated decisions left a big market for Apple to go after and take a big chunk of its future sales and revenue.

Conversely, at HP with mostly non-engineering CEOs at the helm, computer geeks were making continual design use of Microsoft Operating Systems software and OFFICE resulting in the need for protection and additional cost to the user and continual upgrades to ward off new internet and cyber warfare threats. At NASA, in those days long ago, when I spoke of them on new designed systems I was looked at as crazy! They all asked me why NASA of all places would, need such protection on future spacecraft designs, my fellow managers complained the most! As they often say even today, we learn mostly be our mistakes. And as Santana once said, "If we can't learn from our mistakes, we will repeat history over again and again! And as I've shown we as a nation have done just that in both the Challenger and Columbia accidents!

Today, a billion-dollar market has been created addressing software design oversight errors, mistakes and voids. And the threats of growing worldwide cyber-warfare are greater than ever. From my point of view, all this need for extra protection is due to software product design faults and what we used to call "bugs" back in the day.

This new threat by worldwide hostile forces, who today possess in some cases, superior computer software skills that we have in the USA. Our Universities are offering obsolete training in most of their course offerings. Some going back to when I went to graduate school some 40 years ago. Computer curriculums need to be leap frogged to meet today and tomorrow's marketplaces. Cyber warfare and computer security need to have the same vigilance as hardware did years ago. This will take additional checks and balances and Quality Assurance practices many that are like our space and DOD practices that started by Mil-Q-9858 so many years ago.

As I said I own both HP and Apple products and have exclusively used HP since their first one was produced and applied their high-level systems even longer. All my work up to this point was exclusively done using HP products from laptops, desktops and even printers and remote hard drives. However, not only are virus problems and threats a continuous problem, but the complexity involved in just getting a report, memo or letter typed is becoming so cumbersome and complex just to get a document finished in a format one needs to be displayed and readable. We are still not reaching elegant design levels after almost 20 to 30 years of revisions and enhancements. Steve Jobs products closer resemble and reflect a degree of simplicity and elegant design that should serve as benchmarks for the entire high technology business sector.

So, for those new entrepreneurial folks, who have the talent, new world markets will come along filling the present-day voids. The opportunities will be like they were in my 1980s, when I asked my MBA class what were these new industries that would come about in the next ten years and the class and I pondered???

What were they? Nobody could answer…for the answer was right in front of us, but we could not see it…a whole new business called "software" was about to explode that is still in its infancy!

Good Luck and God Speed fellow high technology travelers!

COB

Appendix A: My High-Technology Chronology

Years	Organization/Company and Location
1959	US ARMY Signal Corps School, Fort Monmouth, NJ
1960-61	US ARMY Guided Missile School, Redstone Arsenal, Huntsville, Al
1961-62	Sanders Associates, Geospace Electronics Division, Plainview, LI, NY
1962-64	Lockheed Electronics, Plainfield, NJ
1964-67	Hewlett-Packard, Palo Alto, CA & Boonton, NJ
1967-72	Sanders Associates, Microwave & Data Systems & Federal Systems Divisions, Nashua, NH
1972-73	Infoton, Inc, Burlington, MA
1973-76	Teledyne TAC (Transistor Automation Company), Woburn, MA
1976-80	Kollsman Instrument Company, Merrimac, NH
1980-83	MFE Corporation, Salem, NH
1983-85	Martin Marietta Corporation, Baltimore, MD
1985-88	Martin Marietta Corporation, Maitland & Orlando, FL
1988-88	SMA Corporation, Virginia Beach, VA
1988-88	Texcom, Inc., Lanham, MD

1989-99	NASA Goddard Space Flight Center - Unisys Corporation, Greenbelt, MD
2000-04	NASA Goddard Space Flight Center - SRS Corporation, MD
2004-08	NASA Goddard Space Flight Center - ManTech Corporation, Greenbelt, MD
2008-?	Evans Associates – Senior Technical Advisor, Greenbelt, MD

A few places that I remember teaching....

- Rivier College, Nashua, NH - Undergraduate & Graduate School Programs & Center of Management Development
- New Hampshire College – Undergraduate School, Manchester, NH
- Northern Essex Community College – Haverhill, Massachusetts
- Babson College – Law Enforcement Management Training – Wellesley, Massachusetts
- NH State Police Academy – Management Training - Concord, NH
- Vermont State Police Academy – Management Training - Essex Junction, VT
- Northwestern University – Evanston, IL – Law Enforcement Management Training
- NASA Goddard Spaceflight Center – Flight Assurance Management Part I & II, Greenbelt, MD

In addition, I served on the Planning Board for the new Hollis High School in Hollis, NH and the Catonsville Community College Advisory Board in Catonsville, MD and was a charter founder member of the Center for Management Development at Rivier College in Nashua, NH.

266

Appendix B: My NASA Final Presentation

My ***"NASA's 50 Years in Space"*** Goddard's Contractors Day Presentation to the annual NASA Contractors Convention at GSFC was an opportunity to do my thing…talk to the major players from all over the World about mission success and give a quick overview of how we got to this point in our high technology journey. I have used some of my pitch here to provide that same information for interested readers. As I indicated earlier, this Presentation is available online as well as other informative ones over the past several years. I started talking about the first missile called Matador Program that had 100% failure, until some smart rocket scientist thought about testing and inspecting all the parts used in building the rocket. You guessed it again! After screening they achieved 100% of their launches, thus began the origin of what we have today.

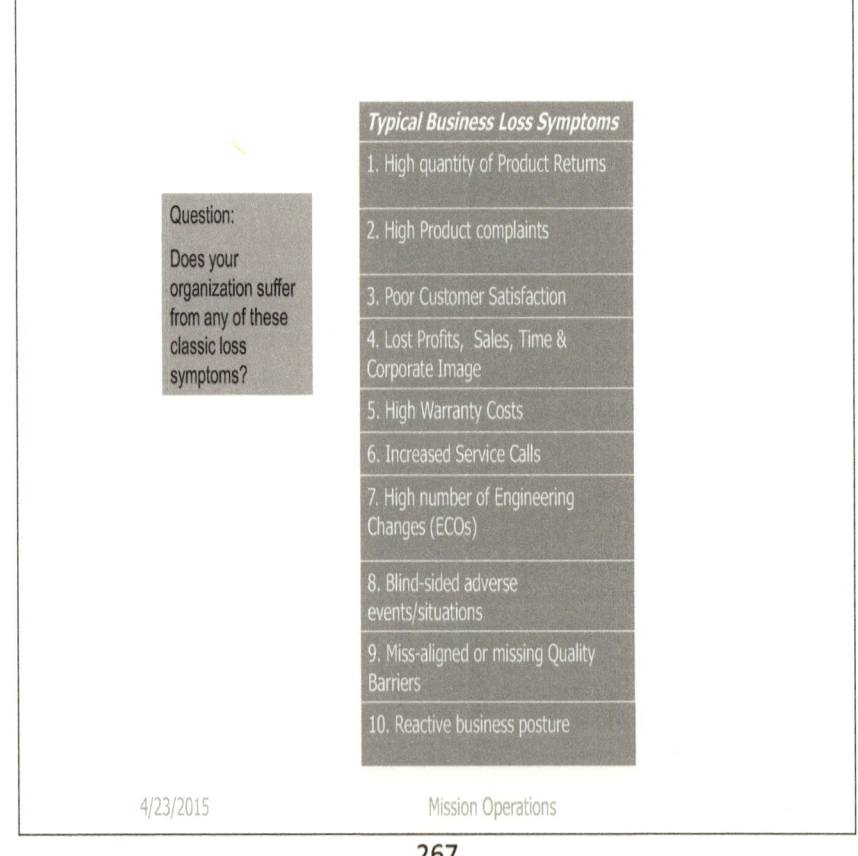

Question:

Does your organization suffer from any of these classic loss symptoms?

Typical Business Loss Symptoms

1. High quantity of Product Returns
2. High Product complaints
3. Poor Customer Satisfaction
4. Lost Profits, Sales, Time & Corporate Image
5. High Warranty Costs
6. Increased Service Calls
7. High number of Engineering Changes (ECOs)
8. Blind-sided adverse events/situations
9. Miss-aligned or missing Quality Barriers
10. Reactive business posture

4/23/2015   Mission Operations

Figure A: My Top Ten List of Typical Business Loss Symptoms

First let us ask a question as shown in the above Figure A. See the list of symptoms and review what may be applicable to your unique business situation. Next let us determine the impact that these various symptoms may have on your business. Figure B illustrates their impact also contains what to do solutions that enable one to take corrective measures to determine root causes, not only the effects, that may not be obvious at first glance! You can see that the business impact can have both short and long-term effects. As I personally found out in many company situations, few firms know how to handle them and are much too late to finally fix them. This is a good reason for having an effective assurance management organization in place; to prevent early warning symptoms from becoming fatal business results.

I also included what we did at GSFC to prevent future on-orbit anomalies from occurring especially in the design, build and test phases of a spacecraft's life cycle. Figure C (see Web page NASASupplierconference2007revA) shows a matrix of causes of anomalies and preventative actions in place to interdict them at the earliest stages in the birth and life of a new satellite. Again, all causes are applicable to any high technology product some that even have been around for over a half-century!

In addition, I covered a seldom used tool for doing preventative accidents, one that became a favorite for its simplicity and effectiveness. What driver doesn't thank the inventor Stevens Institute in New Jersey for those highway life-saving barriers called "Jersey Walls"! Used universally today they are a direct result of applying barrier analysis to a critical application such as the Space program. I went so far as to introduce another layer of safety and quality by using the technique and hoped to influence more of NASA's Contractors to think about the concept in new designs of the future.

Finally, I closed by putting together a picture of how to deal with business problems that plague the industry today from dealing with new threats such as counterfeit parts to new software fault prevention and closing with applying barrier analysis as a measure of rating individual aerospace producer's company performance.

Next illustrates the business loss symptoms, their effects and what can be done to remedy them.

Typical Business Symptoms	Business Impact	What To Do!
1. High quantity of Product Returns	Lost profits	Cause/Effect Analysis
2. High Product complaints	Lost customer confidence	Root Cause Analysis, Pareto Analysis, Barrier Analysis (BA)
3. Poor Customer Satisfaction	Lost sales to competition	Measure, Analyze & Fix
4. Lost Profits, Sales, Time & Corporate Image	High risk producer, High Cost of Sales	Sell real Quality...Stop burning the toast!
5. High Warranty Costs	Profit drainer	Use the 5 Why's (WWWWW)
6. Increased Service Calls	Costly product fixes	Feedback fixes to the Source
7. High number of Engineering Changes (ECOs)	Hidden cost item, band aid fixes	Ship no wine before it's time...Test, Test, Test! (Fly Before Buy!)
8. Blind-sided adverse events/situations	SURPRISE!...When you least expect it!	Know your product limits! Life testing works! (RIDT)
9. Miss-aligned or missing Quality Barriers	Costly Process flaws	Continuously improve the process...forever! USE BA!
10. Reactive business posture	Knee jerk responses	Welcome to the NFL! Practice & Plan your offenses!

Figure B. Business Impact & What to Do!

The Author's Background

**Charles "Chuck" Vincent O'Boyle** has 20 years U.S. Government service (18 at NASA and 2 in the US Army Ordnance Guided Missile School in Huntsville, Al). Prior to NASA, Chuck spent 30 years in industry working for leading high technology organizations in many Engineering and Management roles.

Prior to his NASA experience, Chuck was Quality Assurance Director at Kollsman Instruments, Quality Engineering and Systems Engineering Manager at Martin Marietta, Operations Manager at Teledyne and QA & Reliability Manager at Sanders Associates (ex-Lockheed Martin, now BEA). Early in his career, he worked as a Sr. Engineer for the Hewlett-Packard Company, Palo Alto, CA specializing in microwave instrumentation.

Chuck worked on the Pershing, Patriot, Titan, B1-B, M60 Tank Fire Control System, Tomahawk, Harpoon, USAF F-111 & US NAVY ALQ-126 Electronic Countermeasures Systems (ECM) & Nike Ajax/ Hercules Guided Missile Systems. At NASA, he worked and supervised teams that manned close to 100 successful World-wide launched missions including HST SM # 1 - 3A, JWST, GRO, XTE, EOS AQUA, AURA & TERRA, TRMM, LandSat 7 & the International Space Station (ISS) Fluids & Combustion Facility receiving numerous NASA Achievement Awards for his engineering/management contributions and efforts. The HST Flight Assurance Management Team he managed, including Chuck, won 36 individual NASA Medallions for their work on the highly successful HST Servicing Mission No. 1 that repaired the in-space, orbiting telescope.

For his last several years, prior to retirement in 2008, he worked for ManTech & SRS Corp. as a Lead Engineer. Chuck helped start and supported GSFC Mission Operations for the Office of Mission Assurance covering all GSFC Satellite on-orbit mission performance. He also contributed his expertise in writing the new James Webb Space Telescope Systems Engineering Plan (launch date, now

2021!) and critiqued the James Webb Space Telescope Software Development Plan adding many mission success measures. And he received honors for his role in conducting a worldwide NASA Tracking Station Verification exercise prior to the successful launch and certification of the Earth Orbiting (EOS) AQUA/TERRA Satellite missions.

Chuck taught part-time at the Graduate School level in the Boston area for 10 years and is an Honors Graduate of Northeastern University BS & AS and has a Babson College MBA, with Distinction. He is also a graduate of both Martin Marietta Advanced Management Program (AMP) and Unisys Corporation Advanced Management Program (AMP). And he is a member of Sigma Epsilon Rho National Honor Society, ASQC & INCOSE organizations.

In the 1983, he was given the **Man of the Year** (see photo below) Award by his hometown Friendly Sons of St. Patrick in Pennsylvania along with a key to his hometown city of Pittston at the Annual Dinner Party with hundreds of guests in attendance…an award he is proudest of achieving!

Chuck divides his time today living in Florida in winter (Snowbird) and the Pocono Mountains of Pennsylvania in summer! And he and his lovely wife Julie escape to the heavenly Maine beaches whenever they can!

Hometown Award: *"Man of the Year"* by the Friendly Sons of St. Patrick with Pat Flynn, President on right!

271

Bibliography

Web-Sites:

www.nasa.gov

www.gsfc.nasa.gov

www.hq.nasa.gov

www.ksc.nasa.gov

www.ox.ac.uk - Oxford

www.rivier.edu

www.msfc.nasa.gov

www.jsc.nasa.gov

www.nu.edu

www.babson.edu

www.hp.com

www.apple.com

www.history.army.mil – Packard Commission (1960-2009)

https://www.psu.edu/

https://www.jhuapl.edu/

https://asq.org/

Reaganfoundation.org – Simi Valley, CA Library

Photographs:

Miners, military missiles: Google search engine

NASA Spacecraft & Facilities: Various Nasa web sites (see above)

Huntsville USA Space Museum: Personal collection taken

on visit (https://rocketcenter.com/)

Books & References:

Last Man Out by Glenn McDowell, USMC Amazon Books

FAILURE IS NOT AN OPTION by Gene Krantz Simon & Shuster, NY 2000

The Challenger Launch Decision by Dianne Vaughn, U. of Chicago Press 1996, 2016

Out of the Crises by Dr. Edward Deming , MIT Press Boston 1982

Mission Success in Space by Chuck V. O'Boyle Amazon Press 2019

No BS...Your MBA Primer by Charles V. O'Boyle Amazon Press 2019

50 Years of Mission Assurance by Charles V. O'Boyle NASA GSFC Contractor's Workshop 2007 (Chuck_Oboyle_n_Jim_Suraci_NASASupplierConference 2007RevA)